The

COLLECTOR

The COLLECTOR

DAVID DOUGLAS AND THE NATURAL HISTORY OF THE NORTHWEST

JACK NISBET

Printed in the United States of America
Published by Sasquatch Books
Distributed by PGW/Perseus
15 14 13 12 11 10   9 8 7 6 5 4 3 2 1

First trade paperback edition, 2010

Cover illustration: Jeanne Debons, Douglas fir cone (*Pseudotsuga menziesii*), 2005
Cover design: Rosebud Eustace
Interior design and composition: Sarah Plein/Emily Ford
Interior maps: Jack McMaster
Interior illustrations:
Daniel Macnee, David Douglas portrait, Linnean Society, London (1829)
Thomas Bell, Short-horned lizard (*Phrynosoma douglassii*), *Transactions of the Linnean Society of London* 16 (1833)
William Hooker, Sagebrush buttercup (*Ranunculus glaberrimus*), *Flora Boreali-Americana*, Plate V

Library of Congress Cataloging-in-Publication Data

Nisbet, Jack, 1949-
The collector : David Douglas and the natural history of the Pacific Northwest / Jack Nisbet.
     p. cm.
Includes bibliographical references and index.
ISBN-13: 978-1-57061-613-6 / ISBN-10: 1-57061-613-2 (hardcover)
ISBN-13: 978-1-57061-667-9 / ISBN-10: 1-57061-667-1 (paperback)
1. Douglas, David, 1799-1834. 2. Botanists--Scotland--Biography. 3. Naturalists--Scotland--Biography. 4. Botany--Northwest, Pacific--History--19th century. 5. Natural history--Northwest, Pacific. 6. Scientific expeditions--Northwest, Pacific--History--19th century. 7. Northwest, Pacific--Description and travel. 8. Canada--Description and travel. I. Title.
  QK31.D6N57 2009
  508.795--dc22
                         2009027956

Sasquatch Books
119 South Main Street, Suite 400
Seattle, WA 98104
(206) 467-4300
www.sasquatchbooks.com
custserv@sasquatchbooks.com

CONTENTS

PROLOGUE: NATURE'S HAND ix

I. FRUITS OF THE NEW WORLD: 1823–24 1

II. THE RITES OF NEPTUNE: 1824–25 23

III. BETWEEN THE DESERT AND THE SEA:
SPRING–SUMMER 1825 37

IV. TAKING THE SMOKE:
SUMMER–WINTER 1825 53

V. THE INTERIOR YEAR:
SPRING 1826 73

VI. SLEEPING ON SHATTERED STONES:
SUMMER 1826 91

VII. THE PERFECT ENTHUSIAST:
FALL 1826–SPRING 1827 117

VIII. CROWN OF THE CONTINENT:
SPRING–SUMMER 1827 143

IX. "A SCIENTIFICK NATURALIST":
FALL 1827–FALL 1829 167

X. BREATHING NEW CLIMATES:
FALL 1829–FALL 1832 191

XI. THE CANYON:
WINTER 1832–SUMMER 1833 211

XII. CRATERS: 1834 229

EPILOGUE: NOURISHMENT BEYOND NAMES 251
ACKNOWLEDGMENTS 256
BIBLIOGRAPHY 257
CHAPTER NOTES 265
INDEX 275
MAPS
DAVID DOUGLAS IN NORTH AMERICA: 1824–34 . . . . . vi–vii
THE NORTHWEST COAST: 1825–33 36
THE INTERIOR: 1826–27 72
THE NORTH COUNTRY: 1833 212
HAWAIIAN ISLANDS: 1834 228

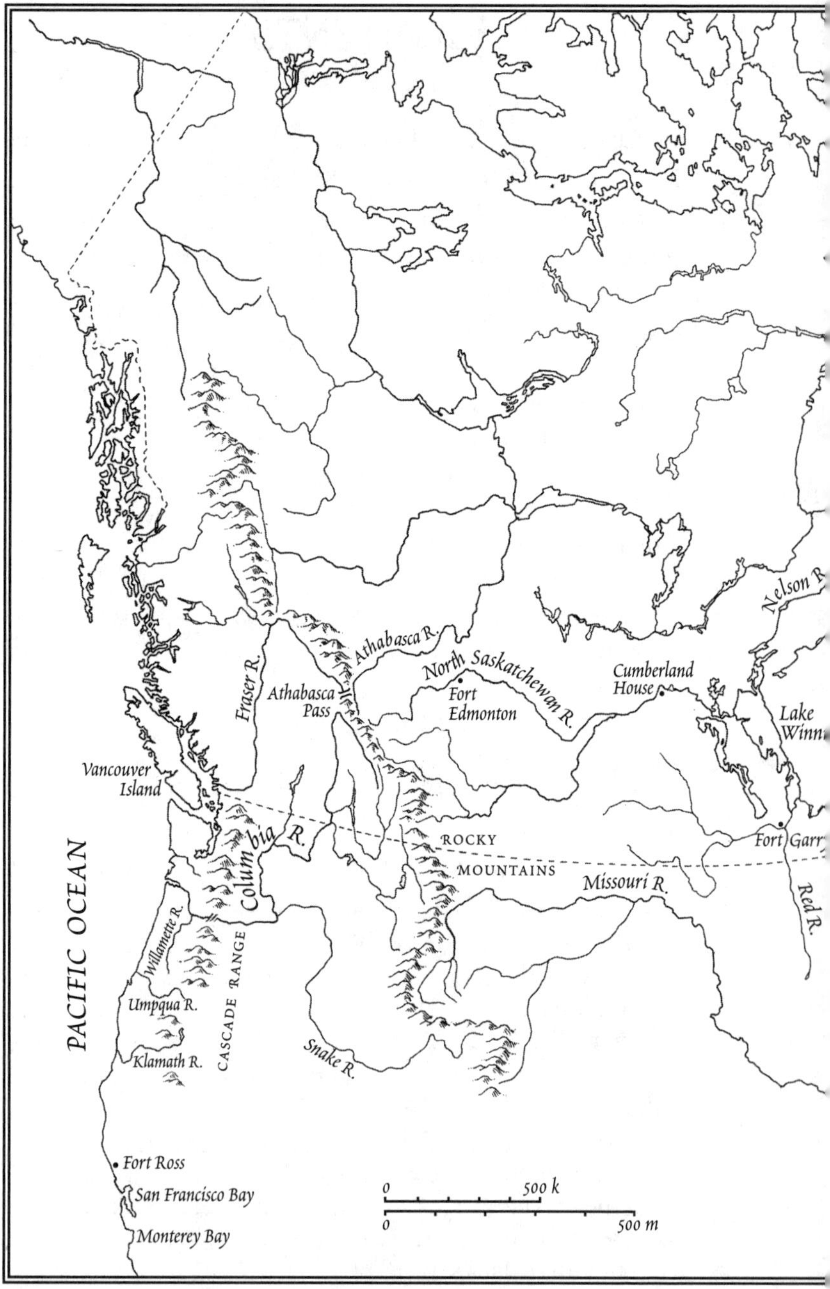

PACIFIC OCEAN
Vancouver Island
Fraser R.
Athabasca Pass
Athabasca R.
North Saskatchewan R.
Fort Edmonton
Cumberland House
Nelson R.
Lake Winn
Fort Garr
Red R.
ROCKY
MOUNTAINS
Missouri R.
Columbia R.
Willamette R.
Umpqua R.
Klamath R.
CASCADE RANGE
Snake R.
Fort Ross
San Francisco Bay
Monterey Bay
0
500 k
0
500 m

David Douglas's Travels in North America, 1824–34

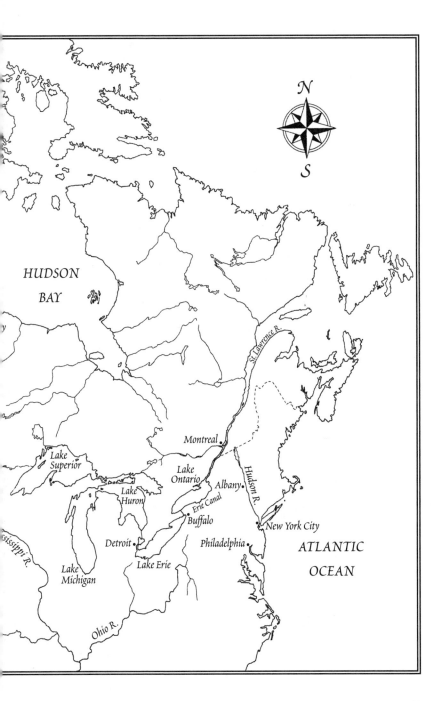

NATURE'S HAND

O n the last day of April, 1827, eight men worked their
way up a snowy slope called the Grand Coté or Big Hill,
the final approach from the Columbia River to Athabasca
Pass. Seven of the group plodded steadily up the steep pitch
toward the Continental Divide, but one of their number, a
sturdy young man wearing spectacles and leggings fashioned
from the sleeves of an old capot, struggled mightily with a set
of unfamiliar footgear.

> Ascending two steps and sometimes sliding back three, the snow-
> shoes kept twisting and throwing the weary traveler down (and I
> speak as I feel) so feeble that lie I must among the snow, like a
> broken-down wagon-horse entangled in his harnessing, weltering
> to rescue myself.

An oblong tin cylinder with rounded edges hung from a
leather strap and bounced against his chest. This collecting

box, or vasculum, identified the straggler as the Scottish plant collector and naturalist, David Douglas. The only member of the party not on the pay list of the Hudson's Bay Company, Douglas was on his way back to England after two years in the Northwest. During that time he had lost neither his self-effacing sense of humor nor his consuming interest in the world around him.

After making camp that evening and gathering dry twigs to start a fire, Douglas melted snow for tea and watched his companions perform their appointed tasks—two or three bringing in firewood, the rest "flooring the house" by piling soft fir branches on the snow. That night, sleeping atop the fragrant cushion with his balding head ensconced in a nightcap woven from the fleece of a mountain goat, he "dreamed of being in Regent Street, London! Yet far distant." Regent Street housed the headquarters of his employers, the London Horticultural Society, who had sent him to this distant place to collect plants. Tucked inside his knapsack, carefully wrapped in oilskin, a tin box held hundreds of small paper packets filled with the seeds of wildflowers and shrubs and trees that would transform the faces of gardens across Great Britain.

Awaking to find himself not in Regent Street but on a snow-covered slope of the Rocky Mountains, Douglas began the day's routine: "rise, shake the blanket, tie it to the top, and then try to see who is to be at the next stage first." When the small party crested the Divide and halted for their midday meal, agent Edward Ermatinger shot a male spruce grouse, the first of this "curious species" that Douglas had seen. He pored over its plumage and crown feathers, "resembling a small well-crested domesticated fowl," then dissected its crop, which was stuffed with black conifer needles. He carefully preserved the bird's skin to take home with him while Ermatinger roasted its flesh. Savoring the succulent meat for lunch, Douglas surveyed his surroundings. Unable to hide his excitement at finding himself upon the great spine of the continent, "I became desirous of ascending one of the peaks, and accordingly I set out alone on snowshoes to that on the left hand or west side, being to all appearance the highest."

The saddle of Athabasca Pass was densely wooded, and Douglas often sank to his waist as he floundered through the corny spring snow. But the toil did not prevent him from plucking samples of a strange club moss to tuck inside the hinged lid of his vasculum. He bent to examine desiccated gentian flowers, and burrowed into drifts to taste frozen crowberries. When he reached timberline, the crustier snow of the open slope allowed him to make much better progress, until the surface abruptly changed to "a mount of pure ice, sealed far over by Nature's hand as a momentous work of Nature's God."

Standing on a shoulder of the peak he later named Mount Brown, Douglas gazed at mountains as far as his eye could see, many of them higher and more rugged than the one on which he stood. Bright sunlight accented the colors of a large ice field below: "the aerial tints of the snow, the heavenly azure of the solid glaciers, the rainbow-like hues of their thin broken fragments." An avalanche tore loose from a south slope to roar away with incredible velocity, "producing a crash and grumbling like the shock of an earthquake, the echo of which resounded in the valley for several minutes." Although he began to wish that he had brought along his flint and steel to start a fire on the chilly heights, his emotions ran closer to ecstasy than to urgency. "The sensation I felt is beyond what I can give utterance to," he later wrote.

He estimated that his ascent had taken five full hours, and the waning afternoon encouraged him to return with all possible speed. Back on the icy crust, he used his stiff snowshoes to great advantage: "Places where the descent was gradual, I tied my shoes together, making them carry me in turn as a sledge . . . sometimes I came down at one spell 500 to 700 feet in the space of one minute and a half." Squeezing every ounce of awe from his toboggan ride down from the top of the world, he found himself back in camp after only an hour and a quarter. He did pay a price for his fun, however, and his journal entry the next morning was distinctly lacking in elevated prose. "May 2. My ankles and knees pained me so much from exertion that my sleep was short and interrupted."

Yet the adventure had surely been worth it. As he lingered on the Continental Divide, Douglas would have been able to look west into the Columbia River drainage, his bailiwick for the past two years. He was well aware of the unique position he had held as he moved through the region's mosaic of communities. To the Hudson's Bay Company, this was the Columbia District, a place to carry on the business of trading for animal pelts. After the expedition of Lewis and Clark two decades earlier, David Douglas was the first visitor with the freedom to focus squarely on the region's natural history, and the first to range beyond the standard travel paths frequented by the fur traders.

He used this opportunity to gather plants at a furious rate, identifying new species with a particular eye for those that showed promise as garden ornamentals. But his interests extended far beyond foliage, and the journals stowed inside his knapsack bulged with descriptions of mammals, birds, fish, rocks, insects, fungi, and anything else that caught his attention. He recorded the activities of humans as well, weaving a portrait of life along the Columbia during the heyday of the fur trade.

When modern naturalists set out to study the flora and fauna of the Northwest, many of the names that roll off their lips and out of their field guides first flowed from the pen of David Douglas. From the mouth of the Columbia River to the crest of Athabasca Pass, he tied many of the first bright ribbons to mark the approach of Western science. Such names only hint at the depth of his accomplishments, however. Douglas's story hurtles through landscapes that only he saw as new, but his finely textured accounts of the Northwest still shimmer against a place that, two centuries later, seems both familiar and drastically changed.

I.
FRUITS OF THE NEW WORLD
1823-24

A Fine Pleasant Breeze

June 3rd—London: left Charing Cross by coach. . . .

T he journal of David Douglas's first great botanical adventure begins like a period novel of intrigue: the hero
steps into a horse-drawn coach that careens through a night
of heavy rain, making for a ship. But instead of heading east
toward a clandestine rendezvous at the Channel, Douglas's
route led west, to Liverpool. Rather than scanning the faces
of men for clues at a stop along the way, the young detective
peered into a horse trough, where he ascertained the true
identity of a particular green algae whose branches danced
across the water's surface.

Back on the road, the coach creased a landscape that
varied from wild to artfully arranged. Instead of pondering
the woman across the carriage, Douglas turned his gaze to
the beautiful fields around Woburn Abbey, "tastefully laid
out and divided by hedgerows in which are planted Horse-
Chestnuts at regular distances, all in full flower." As more

romantic scenery swept past, an aquatic plant scattered across a natural wetland caught his eye; he carefully penned its Latin species name, *nymphoides*, marking its place in the great lattice of botanical taxonomy.

Late in the afternoon of the second cold, rainy day, the coach pulled into Liverpool, where Douglas learned that his ship was set to depart the following morning. That allowed time to nip by the Botanic Garden, where a certain Mr. Shepherd "received me in the most handsome manner, and showed me all his treasures." More than a few of Shepherd's jewels were from North America, and Douglas studied them eagerly in preparation for his own journey to that promised land. He thought the ferns particularly pleasing, but an array of pitcher plants somewhat weak. With the eye of a true student, he scrutinized containers, soil mixture, and the angle of exposure for species gathered from around the world.

Too soon, the budding traveler was obliged to return to the dock. When the tide proved too low for the ship to clear the River Mersey, he hopped ashore and made his way back to the gardens to marvel at the woolliness of a bromeliad from Trinidad. Plants imported from the tropics, sheltered within heated greenhouses, were objects of great curiosity in the northern climes, and he admired several other Caribbean species, located by the practical Mr. Shepherd against the back wall of a stove.

The following morning, Douglas stood on deck as two steamboats towed the *Ann Marie* down the channel. Contrary winds again stymied her advance toward the New World, and she spent the entire next day tossing in place. While most of the passengers suffered greatly from seasickness, Douglas "felt perfectly comfortable, only a headache which was occasioned by cold when on my way to Liverpool." As a steady gale replaced the blustery crosswinds, he wrote off the stormy conditions as "a fine pleasant breeze." Upon reaching the wide Atlantic, he faithfully recorded the latitude taken each day at noon by the navigator. On June 26, 1823, twenty days out of Liverpool, he marked the beginning of his twenty-fourth year on the planet: "This being my birthday and the market day of my native place, I could not help thinking over the days that were gone. Light airs of wind; making little progress."

The Apprentice

In contrast to the *Ann Marie*'s slow advance across the North Atlantic, David Douglas had made a great deal of progress since his birth in June, 1799 in the village of Old Scone, just north of Perth, in east-central Scotland. His father was a stonemason who engaged in specialty work such as tombstone statuary and firebox repair. Although limited in income, the elder Douglas contrived to educate all six of his children; David, the second of three sons, began school when he was about three years old. From the beginning, his natural high spirits, combined with a certain obstinacy, resulted in frequent clashes with his teacher, "a worthy old lady" from the village.

At around age seven he was enrolled in a school in a neighboring parish that was run by a man known as an inflexible disciplinarian. The two-mile walk to this school presented many opportunities for a boy who disliked the restraints of the classroom to dally among copses and heather-covered hills, and Douglas soon distinguished himself for tardiness and truancy, and by "his contempt for the master's thong, and his carelessness about those difficulties and hardships which would have weighed hard with other boys." According to his elder brother, John, the corporal punishment did not faze young David, whereas being held after school was the severest penalty imaginable.

Possessed of a quick intelligence and excellent memory, he became proficient in Latin, penmanship, and basic arithmetic. From an early age, natural history consumed his interest, with his favorite occupations being fishing, picking wildflowers, collecting birds' nests, and capturing live birds. "I remember him once finding a nest of young owls which he managed to keep very well for a short time by means of a mouse trap," John later recalled. When the mice became scarce, David traded his lunch penny for bullock's liver to feed the owlets, then either begged half his brother's lunch or went without. He became an expert fisherman, and "when he could not procure proper tackle he had recourse to the simple means of a willow, string, and crooked pin with which he was often successful." John openly admired his younger sibling's inquisitive nature, giving as an example an incident that occurred one day in Perth, when the brothers passed a shop where a man was preparing rough tobacco leaves for sale. The process so captivated David's attention "that he could not rest satisfied until he had learned its whole history."

When David was about eleven years old, he began working during the summer for family friend William Beattie, head gardener at Scone Palace, the local manor. Although the boy's capacity for hard work impressed Beattie, his penchant for mischief and "quarrelsomeness" created friction with other lads working on the grounds. But when they complained to their boss, Beattie supposedly replied, "I like a devil better than a dolt."

During the next few summers, David continued to work in the palace nursery and flower gardens, becoming a great favorite of Beattie and his assistant. John Douglas believed "it was in this place where he acquired that taste for botanical pursuits which he so arduously pursued in after life." In 1815, when David was sixteen, he was apprenticed to Mr. Beattie and assigned to the kitchen gardens. The hours were long, but in his spare time Douglas made excursions to the surrounding woods and hills, where he collected plants for a small garden he tended at home.

Despite his earlier scholarly reluctance, he was in no way averse to knowledge, perusing every available work on natural history and travel. One of his close childhood friends, William Beattie's nephew and namesake, William Beattie Booth, observed that on winter evenings David could invariably be found absorbed in books he had borrowed from friends and fellow workers, "making extracts from them of portions which took his fancy, and which he would afterwards commit to memory." His brother John believed that his voracious reading of travel narratives and popular adventure stories such as *Robinson Crusoe* and *Sinbad the Sailor* inspired an interest in the world beyond Scone. Booth, however, did not expect his friend's interest in adventure to translate into a career. "Whether he had any intention of becoming a Botanical traveler we have now no means of ascertaining, but we are inclined to think not, for he was then so timid, as scarcely to be able to go any where alone after nightfall; and was altogether destitute of that daring intrepidity which he subsequently displayed."

After completing the terms of his apprenticeship, Douglas spent the winter of 1817–18 in a Perth school known for its mathematics curriculum. Meanwhile, Beattie secured a position for his former pupil as assistant gardener on Sir Robert Preston's estate at Valleyfield, about twenty-five miles to the south. There Douglas was introduced to the latest fashions of

Scottish gardening, including the cultivation of exotic ornamentals, and was allowed the run of a well-stocked library.
The young man's spirit and curiosity impressed Preston and
his head gardener: "Douglas was so enthusiastic that he would
see more in anything he took to than what the generality
of people could perceive." The following year, they recommended their lad for a job at the botanic gardens of Glasgow
University.

Douglas moved to the smoky industrial city of Glasgow
in the spring of 1820, followed one semester later by the col-
lege's new garden director and professor of botany, William
Jackson Hooker. This thirty-five-year-old Englishman was on
fire with the wonders of the natural world, and although he
had never conducted a single class before this appointment,
he proved to be a charismatic teacher. His popular lectures to
beginning medical students were held in a small building in
the center of the gardens, and Douglas sought permission to
attend, greatly enriching his knowledge of plant physiology
and taxonomy. Hooker, an accomplished artist, hung large
colored drawings of plant organs around the classroom and
brought in piles of dried specimens for his students to analyze.
Excited by all that he was learning, Douglas sometimes forgot
the formal conventions of nineteenth century academia, and
"once rushed into the lecture room, just in his shirt and trousers, as he was working, with what he considered a new and
rare plant for Sir William to look at."

Hooker began inviting the young nurseryman to join his
students on botanizing excursions around Glasgow, where
Douglas honed his field collecting skills. On longer summer
expeditions in the Highlands, the professor and his protege
found that they shared a droll sense of humor as well as a
love of hunting and fishing. Welcomed by the affable Maria

Hooker to her bustling household, Douglas became a great favorite of the couple's young children, especially the two boys, Will and Joseph, who often trailed their father to the university gardens and on rambles in the country.

In the professor's extensive herbarium (housed in the family laundry), Douglas could practice the delicate craft of preparing dried specimens. In this process, which he called "laying in," fresh plants were carefully arranged on sheets of fine white paper, which were then separated by thicker rag blotters and squeezed between the boards of a large plant press. While meticulously labeling each sample by genus, species, and location, he also would have learned a great deal about the evolving science of classification, for Hooker was an influential force in developing a "natural" system for ordering the plant kingdom, refining the binomial scheme pioneered by the Swedish taxonomist Carl Linnaeus.

Three years of acquaintance impressed Hooker with Douglas's "great activity, undaunted courage, singular abstemiousness and energetic zeal." When the London Horticultural Society put out a call for a botanical collector in the spring of 1823, the professor recommended his disciple as "an individual eminently calculated to do himself credit as a scientific traveler."

Some might not have considered this appointment much of a favor—collectors often endured harsh conditions, as witnessed by the recent deaths of two of the Society's employees: one from fever in East Africa and another from an illness contracted on assignment in India. Nor was there great prestige associated with the position, as Sir Joseph Banks, one of the founders of the Horticultural Society, had made clear when he set down some precepts on the subject: "Collectors must be directed by their instructions not to take upon themselves the character of gentlemen, but to establish themselves in point of board and lodging as servants ought to do." But there were men of modest standing who had reached great heights, such as Robert Brown, the "Jupiter" of British botanists, who had begun his career as a traveling collector.

In Great Britain at that time, botany reigned as the most important of the natural sciences—the basis of all medical training, and an essential cog in the economic engine of the Empire. Naval officials eagerly sought new sources of timber for shipbuilding, and crops with commercial potential, from breadfruit to tobacco and cacao to corn, were transplanted from one colony to another like companies of militia. In addition to practical and scientific interests, the importation of exotic plants had become such a favorite hobby of the upper classes that "distinguished persons would throw themselves almost into a frenzy when the rumour went abroad that some new or rare specimen was about to appear on the market."

In 1804 an eclectic group of enthusiasts founded the Horticultural Society of London, dedicated to advancing scientific botany, practical horticulture, and ornamental gardening. The Society progressed slowly at first, but within two decades, honorary secretary Joseph Sabine had developed the organization's social and political connections into over a thousand contributors, including sixty foreign correspondents who sent seeds and cuttings of unusual plants from faraway climes. With the tsar of Russia and the kings of Denmark, Bavaria, and the Netherlands on his rolls, Sabine could be sure that doors would open for anyone who traveled under Society auspices.

By 1820, Sabine had engineered the purchase of thirty-three acres in the London suburb of Chiswick for an ambitious botanical garden. The layout was designed by architect

William Atkinson, who had overseen the reconstruction of Scone Palace and its grounds (where he hired John Douglas as an apprentice and became close friends with the entire family). For the Chiswick gardens, Atkinson chose a rambling Rococo style, "with the species of each genus placed in the immediate vicinity of each other, thus affording an opportunity of convenient comparative examination."

The Society was devoting one portion of their new acreage to exotic plants from around the world, and they initially planned to send their new collector to Chile, but political concerns squashed that mission. "In order not to leave Mr. Douglas unemployed," the secretary reported, "it was conceived that he might be most usefully engaged in bringing from the United States such plants as were wanting in our collection, particularly fruit trees." Douglas was to investigate the latest developments in American fruit-growing and obtain new apple and pear varieties for an experimental orchard at Chiswick. He was also to acquire samples of forest trees, particularly oaks that might help replenish the dwindling homeland timber supply. Finally, in his spare time, Sabine encouraged him to botanize for wild plants that might be naturalized in English gardens.

Predictable Vigor

As the *Ann Marie* wallowed across the North Atlantic, the new collector perused scientific journals, practiced Spanish grammar (Joseph Sabine had suggested that the language might come in handy someday), and studied the seagoing culture. He watched crewmen nurse their last bits of tobacco by laying chewed plugs out to dry, then smoke the remnants in their pipes. The captain had rationed drinking water, and in the aftermath of a heavy rain shower, Douglas "could

not help but observe how the dogs eagerly licked the decks."
Thirsty for some action himself, he welcomed the sight of
George Bank with its myriad birds; the tight clusters of fish-
ing boats; the taste of fresh mackerel; the spectacle of Cape
Cod in the distance; and finally, on the last day of July, "Long
Island in sight; I cannot express the satisfaction I feel; shores
sand and partly rock."

After two months at sea, Douglas was eager to get ashore,
but the city's health officer announced a two-week smallpox
quarantine that limited passengers to brief visits on Staten
Island. Not until August 6 did Douglas finally receive permission to visit the city, and then only on the condition that he
buy new clothes and leave his old ones behind, lest they harbor
infection. That day he stepped ashore to a fresh new world.

> This morning can never be effaced; it had rained a little during the
> night, which cooled the atmosphere and added a hue to Nature's
> work, which was truly grand—the fine orchards of Long Island on
> the one side, and the variety of soil and vegetation of Staten on the
> other. I once more thought myself happy.

His state of mind improved even more when he interviewed Dr. David Hosack, one of Joseph Sabine's most
active contacts. A prominent physician who had attended
Alexander Hamilton after his duel with Aaron Burr, Hosack
was also a respected naturalist, happy to exchange ideas with
professionals and amateurs alike. The son of a Scottish shop-
keeper, he had studied medicine in Edinburgh and apparently
felt a personal affinity for the young Douglas, inviting him to
his home and alerting a network of acquaintances throughout the state to welcome the visitor.

After this cordial reception, Douglas called on one of
the doctor's former students, botanist John Torrey, who was

preparing an important flora of the Mid-Atlantic region and impressed Douglas as most agreeable and "much disposed to aid me." William Hooker had sent along a book on mosses and a few dried plants as gifts, and Douglas apparently explained how his professor had classified one of the specimens. Torrey, writing to Hooker about the matter, inquired: "Is it true as that fellow Douglas says (he is such a liar I know not whether to believe him or not) that it [pinedrops] is allied to Pyrola [wintergreen]?" Douglas was correct here—pinedrops and wintergreen are both members of the heath family, but for some reason Torrey reacted to this information with impatience. He had a reputation for irascibility, and on one occasion had accused even his good friend Hosack of lying. Douglas, either unaware of Torrey's bile or unfazed by it, wrote to Hooker only of the "kindness and attention" he received.

Douglas spent the next few August days touring the city with a more even-tempered companion, an English acquaintance of Dr. Hosack named Thomas Hogg, who ran a nursery and florist business on Manhattan Island's Broadway. During a joint excursion to the famous Fulton vegetable market, Douglas cast a kitchen gardener's critical eye on every booth they passed. The beets, carrots, and red onions he deemed very fine indeed, but he fretted over a "very great deficiency of cauliflower." Immense mounds of melons met his approval, but he was sorry to see that the cucumbers were the same short prickly ones he knew from home. He did find new fruit varieties in abundance: apples, pears, peaches ("picked far too green"), and some early damson plums.

After their market inspection, Douglas and Hogg set off for a commercial nursery in Flushing owned by William Prince, another correspondent of the London Horticultural

Society. Prince had every reason to believe that he would be selling many plants to the Society's representative, but Douglas was not impressed: "On the whole I must confess to be somewhat disappointed, for his extensive catalogue and some talk had heightened my idea of it; but most of his ground is covered with weeds." Douglas dismissed the grower as "a man of but moderate liberality" and resolved to look for other suppliers of rooted stock.

The next day, he and Hogg crossed the Hudson River to survey some of the nearby orchards. In conversation with an elderly Dutchman who tended twenty-four varieties of peaches, Douglas recorded details of cultivation ranging from row and seed intervals to weeding and pruning techniques. The Dutchman's farm bordered a swamp teeming with pitcher plants and club mosses, and Douglas dug several to tuck into his vasculum, certain that they held great promise as ornamentals.

Soon after returning to his lodgings, Douglas received a package from Dr. Hosack containing four plums of the special "Washington" variety—"a name which every product in the United States that is great or good is called," Douglas commented. The delicious yellow fruits, which looked rather like greengages with thin skins and very small pits, had an intriguing history: "Purchased by a Mrs. Miller about thirty years since out of the flower-market. After standing in her garden for five years it was during a thunderstorm cleft nearly to the bottom, which caused its death so far as was rent." The following spring, a local orchardist noticed suckers sprouting from the scarred trunk, and planted one in his own garden. A few years later, it produced fruit without any grafting. He pampered the young tree with rich soil and compost, noting that the fruit's flavor improved year by year. Each winter he laid the roots bare, then replaced the soil in the spring. This was exactly the type of maverick stock and specialized knowledge that the Society members were hoping to import from America. The four Washington plums that Douglas carefully suspended in a jar of alcohol epitomized the purpose of his journey: to capture some of the New World's unpredictable vigor and infuse it back into the Old.

For the next four months, Douglas continued an energetic swirl of activity. In mid-August, he and Thomas Hogg set off for Philadelphia, where they interviewed nurserymen and visited various gardens and landmarks, including the natural history exhibits at Peale's Museum, where Douglas was quite taken with the broad curved horns of a mounted bighorn sheep that Lewis and Clark had shot on the Missouri River in 1806. At the University of Pennsylvania, they met William Dick, a janitor whose duties included caring for the school's small botanic garden. Dick had succeeded in propagating seeds from recent expeditions to the Arkansas River and the Rocky Mountains, and he generously offered to share cuttings with his visitor.   .

The search for superior nursery stock next led Hogg and Douglas south into Delaware, "said to be good for the neatness of its gardens," but the travelers saw nothing particularly noteworthy until a roadside ditch presented an array of aquatic plants, including the famed American lotus. Douglas would later try to dig one of these plants, but its roots were too deeply embedded in the mud.

Back in New York City, the men carried their gleanings to Hogg's garden, where Douglas heeled them in with the assistance of Hogg's son, who "carefully attends to the little treasures during my absence." In the interest of accumulating more treasures, the collector attended a meeting of the New

York Horticultural Society, whose members offered every kind of assistance. Ever more attuned to the advantages of local knowledge and influence, Douglas visited several of the committee members before setting off on another excursion. September 4 found him steaming beneath the palisades on the Hudson River, drinking in the lore of West Point as he cruised past. That evening he was welcomed by the family at a nearby estate, where he was excited to find four large oak trees, of four different species, whose acorns he made plans to gather on his way back to the city.

Upon arriving in Albany, he presented himself to Governor DeWitt Clinton, a close friend and former class-mate of Dr. Hosack. Clinton heartily advised the young collector to head for Canada straightaway so that he could witness the full glories of autumn in the eastern woodlands. The obedient Douglas accepted "letters of introduction to all the places of science or influence in my line of journey," then hopped aboard a 4 a.m. stagecoach to endure seventy miles of jolting on a rough corduroy road along the Mohawk River. Abandoning that uncomfortable mode of travel, he embarked on the Erie Canal, the governor's signature proj-ect. As the canal boat pulled out of town, it passed beneath the triple granite arches of an elegant bridge, dedicated to His Excellency DeWitt Clinton.

From Utica to Rome to Rochester, Douglas floated in comfort, gazing at rich hardwood swamps along the way. After a night coach to Buffalo and sixty hours aboard a steamer that ran the length of Lake Erie, he reached the mouth of the Detroit River, where the tongue of Ontario licks beneath Michigan's rising shoulder. A political flashpoint since the earliest French and English border skirmishes, the area remained disputed territory between Canada and the United

States. To circumvent restrictions governing American passengers, Douglas was set ashore on a small island in the Detroit River on September 15. An Indian in a birchbark canoe ferried him to the British garrison near the village of Amherstburg, Ontario, where he called on Henry Briscoe, an officer who had served with Joseph Sabine's brother Edward during the War of 1812. Happy to hear news of his old friend, Briscoe entertained his guest with conversation about sporting birds, Indian corn, and cultivated tobacco, then offered to take him on an excursion first thing the next morning.

The First Real Day

The men set out with dogs by 6 a.m.; within two hours Briscoe had proved himself to be a crack shot, and Douglas could only remark, "This is what I might term my first day in America." The dark humus of the forest floor nurtured wild oaks, hickories, butternuts, and black walnuts, "some of immense magnitude." In a drier situation he found two cancer-roots, odd saprophytic plants that grow without chlorophyll beneath beech and oak trees. If he could keep them alive, they would make fine curiosities.

Traveling along the Detroit River during the next few days, the collector compared the techniques of French fruit growers with those he had seen in upstate New York. The peaches, he judged, "have not that sickly appearance which they have in the States, occasioned probably by the excess of heat." He explored nearby woodlands, walking tirelessly for hours. He filled his vasculum with live plants, seeds, and cones. As with many enthusiasts, his focus on certain objects of desire bordered on the obsessive. His heart fluttered whenever he passed a swamp that could nurture carnivorous pitcher plants, or heavy duff that he suspected might harbor saprophytic specialties. The oak family, so important to Sabine, always remained uppermost in his mind, and he harvested acorns at every opportunity, keeping a careful record of their locations. He gathered favorite species of his mentors back home, such as honeysuckles and a fine ladyslipper orchid. "I wish they may grow," he wrote in his journal.

Briscoe recommended a trip to Lake St. Clair, so Douglas pushed north to the village of Sandwich. On September 20, he hired a buggy and driver to conduct him farther along the lake. When the horse became exhausted from trudging through swamps and dunes, Douglas left the animal to recover while he and his attendant struck off on foot. After walking about five miles, pocketing many new plants along the way, Douglas spotted a promising stand of oaks. Pausing beneath one enticing tree—number 43 on his list of oaks he had sampled—he decided he would have to climb the tree in order to secure prime acorns. The day being warm, he discarded his coat and ascended the stately branches.

> I had not been above five minutes up, when to my surprise the man whom I hired as guide and assistant took up my coat and made off as fast as he could run with it. I descended almost headlong and followed, but before I could make near him he escaped in the wood.

Upon giving up the chase, Douglas realized that the pockets of his coat had contained not only his plant notes of the day but also some valuable receipts, nineteen dollars in paper money, a copy of Persoon's *Synopsis Plantarum*, and his small vasculum. Seething with righteous anger, he tied the acorns from Oak #43 in his neckerchief and trudged back to his buggy. There, however, "the horse only understanding the French Language, and I could not talk to him in his tongue, placed me in an awkward situation."

Though irked by the loss of his specimens and jacket, Douglas did not allow the misfortune to deter him from the work at hand. After searching in vain for a tailor to make a new jacket, he borrowed a spare from an acquaintance and returned to Amherstburg. Crossing the Detroit River for a day's exploration in the Michigan Territory, he tangled with greenbriers in a dense forest and plucked a tiny moonwort from among the dry leaves. He searched without success for trillium but did come upon an odd rose that had been brought from France by an early settler to decorate a garden and then escaped into the wild.

On September 23, he learned that Briscoe was being reassigned to an eastern garrison. After only ten days in Canada, the collector was far from satisfied with his gains, but a spell of foul weather and his host's imminent departure convinced him to head back to New York. Two days later, he boarded the Lake Erie steamer with the Briscoe family. A violent squall ripped off a paddle wheel, but they docked safely in Buffalo, then braved snow showers for a visit to Niagara Falls. Douglas, "sensitively impressed with their grandeur," was soon wandering through the mists of Goat Island, picking up limestone and gypsum samples and plucking more moonworts and cancer-roots.

After bidding farewell to the Briscoes, Douglas traveled a short distance to Lockport to meet David Thomas, another amateur naturalist recommended by Governor Clinton. At the moment Thomas was a bit of a hero, for he served as chief engineer for the just-completed western district of the Erie Canal. As he toured Douglas along the five great locks leading to Lake Erie, the engineer revealed a predilection for birds and minerals as well as for stonework, and the next day he led his visitor through a beech woodland abounding with orchids of all sorts, from wonderful coralroots and pogonias to ladyslippers and showy orchis.

Douglas bounced back to Rochester, then again found respite in a canal boat as far as Utica. Tucked inside a coach for the final leg to Albany, he became chilled during the night and "was seized with a rheumatism in my knees, which alarmed me a little." In the other parts of his journey, Douglas had endured harsh weather many times, but this was his first mention of any physical effects. The joint pain had eased by the time he stepped from the coach in the capital on October 9, where he found celebrations under way for the opening of the western section of the canal: "The town was all in an uproar—firing of guns, music, &c. . . . I had considerable difficulty getting lodgings in the inn."

The next morning he breakfasted with Dr. Hosack, who was in town for the festivities, and together they called on the governor. Then Douglas excused himself for some important sleuthing. David Thomas had described a likely spot south of town to look for saprophytic pinedrops, and after searching for seven hours, the collector located not one but two populations, growing in entirely different situations. The next day he returned to the site and dug up every one of the plants that he could find. His avidity did not go unnoticed—a visitor to New York the following year learned that some of the state's plant lovers were "furious against Douglas for having deprived the Flora of New York of several rare plants which he carried off root and branch from the only localities where they are known to exist."

A Gift of Seeds and Seedlings

Although Douglas had originally planned to sail for London in October, Governor Clinton and Dr. Hosack convinced him to tarry a little longer. Glad to have extra time, Douglas returned to Philadelphia and paid another visit to the University of Pennsylvania, where he was selecting samples from William Dick's extensive seed collection when he had the good fortune to meet Thomas Nuttall, the dean of American naturalists and author of a plant manual that Douglas consulted daily during his travels. Born of ordinary means in Yorkshire, Nuttall had sailed to Philadelphia at age twenty-two. Beginning with minerals, he had mastered every facet of American natural history, collecting flora and fauna up and down the Eastern seaboard; after a difficult trek along the Arkansas River in 1818–19, he completed his landmark *Genera of American Plants* in 1822. Reflecting both his lack of European training and the new wave of scientific ideals, Nuttall wrote his plant descriptions in English rather than Latin, but that eccentricity did not prevent his book from being proclaimed a universal success.

Douglas found Nuttall to be a gracious host, and on November 3 they traveled together to the famed Bartram garden on the Schuylkill River, where John Bartram and his son William had been nurturing the New World's botanical ardor since the 1750s. A beloved mentor to Nuttall, the elderly William Bartram had passed away a few months before, but his niece, Nancy Carr, was there to greet them, and Douglas learned that she too was a "considerable botanist and draws well." Either Mrs. Carr or Nuttall pointed out a large cypress tree in front of the Bartram house, which had been planted with great ceremony by John Bartram himself: "His son William (the late) held the tree while his father put the earth round; it is eighty-five years old."

During the years before the American Revolution, William Bartram had made a series of collecting journeys through the

Southeast. His narrative of those tours, combined with his bird knowledge and ethereal watercolors, had gained him international respect, and Douglas moved about his gardens with reverence. On the banks of one of the ponds grew a large sourwood tree, known for its nectar-rich blossoms. As he collected a packet of its seeds for propagation back in London, he learned that Bartram had tried for more than forty years to sprout progeny from this special tree, but never succeeded till the previous spring, "when he had the gratification of transplanting an abundance of them in small boxes two days before he died." Mrs. Carr bequeathed a dozen of the seedlings to Douglas on the condition that he share them with a London nurseryman of her uncle's acquaintance.

Visiting the hallowed garden of William Bartram in the company of Thomas Nuttall would have been an especially memorable occasion for the young Douglas. Between them, the two Philadelphians exemplified a combination of wilderness exploration, careful scholarship, and poetic storytelling that an ambitious young man like David Douglas could hope to emulate. And, as their travels made clear, there remained vast stretches of North America that brimmed with a bounty of flora and fauna yet to be described.

Upon his return to New York City on November 6, Douglas enjoyed his customary Sunday breakfast with Dr. Hosack, then boated across the Hudson with Hogg to search for pitcher plants. "After some difficulty and all besmeared with filth, reached them, having to carry them two miles through the swamp. Darkness put a stop to our pursuit before we could get enough plants."

The muddy collector spent the rest of November preparing for his departure. He sorted incoming shipments from nurseries he had visited, ordered crates, packed plants and root stock, and ushered box after box to the wharf. On December 5 he made one last visit to the "infamous" Mr. Prince in Flushing, from whom he had purchased only plants that he could not find elsewhere at a better price. "Our words were not of the most amicable tenor," he wrote, "and I am sorry to say that I must leave America without having good feelings towards every person."

Other farewells were far more congenial. After making sure that jugs of apple cider and cages of wood ducks, quail, and pigeons (gifts from various New Yorkers to acquaintances in London) were safely stowed aboard the vessel *Nimrod* on December 9, he spent the afternoon with Dr. Hosack and "a large party of friends who kept mirth up till a late hour in the evening." While Douglas was attending yet another farewell dinner the next night, the *Nimrod* hauled out into the harbor, and Thomas Hogg performed one more favor by hiring a rowboat to transport the tardy collector to his ship.

On the voyage across the Atlantic, Douglas kept a close eye on the birds entrusted to his care. The ducks became very seasick and ate nothing for two days; Governor Clinton's prize pigeons tussled with another pair that had "most stupidly" been put in the same cage. "They fought furiously and four fell in the engagement, two on each side," he reported to Clinton, adding that the combatants all soon recovered. After some slow going in the Gulf Stream, the *Nimrod* caught a favorable wind, and on New Year's Day 1824, she hailed the rocky shores of Cornwall.

II.
THE RITES OF NEPTUNE
1824-25

Success Beyond Expectation

A s the *Nimrod* dropped anchor in the Thames on January 10, 1824, Douglas wrote that he "had the pleasure of arriving safe at London on Friday morning having had a highly interesting journey," then closed his journal. Although no record of his activities for the next several days survives, he must have been occupied with transporting bird cages, cider jugs, and boxed plants to the Society's garden in Chiswick. Upon opening one of the cartons, he found that some of his wintergreen and trailing arbutus were in flower—stimulated, he concluded, by the warmth of the Gulf Stream. As other healthy specimens emerged from the crates, it became clear that Douglas had a special talent for procuring and packing live plants and healthy rootstock.

Shortly after his return, the Society's *Transactions* contained a glowing announcement: "His mission was executed by Mr. Douglas with a success beyond expectation; he obtained many plants which were much wanted, and greatly increased our collection of fruit trees by the acquiring of several sorts known only to us by name." In addition to new peach and apple varieties, a Washington plum from Dr. Hosack, and a much-desired cutting from the famous Stuyvesant pear, his offerings of other American flora soon became popular items in the Society's nursery. One of the honeysuckles to which he had devoted special care quickly sprouted fresh growth in a hothouse. An Oregon grape, propagated from seeds harvested by Lewis and Clark (and probably obtained from the contrary Mr. Prince), had survived the transatlantic voyage and soon thrived in the English climate.

The unusual plants that Douglas had obtained from generous American gardeners "embraced nearly everything which it was desirous to obtain," Sabine gushed. Up in Glasgow, Professor Hooker learned that his former student had "done himself great credit, and given his employers the highest satisfaction, during his mission to the United States." DeWitt Clinton weighed in from across the Atlantic, writing that his young British visitor "unites enthusiasm, intelligence and persevering activity" and "appears to me to combine the essential qualities required in trusts of this nature."

Not all of Douglas's encounters proceeded so smoothly, however. The Society's president, Sir Thomas Knight, an orchardist of great repute who naturally looked forward to conversing with the collector about his experiences abroad, found the young horticulturist "the shyest being almost that I ever saw . . . until I had talked to him for some time in a friendly and familiar way." Many adjectives would be applied to Douglas during his lifetime, but rarely was the description "shy" among them.

Perhaps Douglas was weary of socializing and eager to get to work, for he had much to do. A multitude of pressed plants waited to be cataloged; seeds needed to be sorted and sprouted; his field notes had to be compiled. In addition, he had agreed to organize a collection of dried plants donated by John Richardson, the naturalist on Sir John Franklin's first Arctic expedition. Joseph Sabine asked Douglas to assemble a set of duplicates for an American herbarium in recognition of the generous donations from so many New Yorkers and Philadelphians to the Society.

As if all this was not enough, Douglas was writing a paper on the American members of the genus *Quercus*. The surviving manuscript bears the title "Some Account of the American Oaks, Particularly of Such Species As Were Met with During a Journey From * to * in the Year 1823 by David Douglas." The author offered an account of the species classifications of that time, followed by common names and varying amounts of discussion. He had thoroughly researched the available literature, collating comments from pioneering botanists Frederick Pursh, Thomas Nuttall, and especially Andre Michaux. The junior collector added a few of his own observations, including an opinion that wood of the red oak "is of very inferior quality, scarcely fit for any purpose but the making of barrels."

Douglas had been back in London for six months when Joseph Sabine received a letter from the Hudson's Bay Company in late June 1824 offering free passage for a collector to work in the Pacific Northwest. Apparently undeterred by the fur company's caution that a visitor might "find the fare of the country rather coarse and be subject to some privations," the Horticultural Society immediately nominated Douglas for the job.

The Bay Company's annual supply ship was scheduled to depart in late July, and Douglas prepared as thoroughly as possible in the short time available. He visited Archibald

Menzies, who had served as physician and naturalist on voyages to Alaska in the late 1780s and with George Vancouver's Pacific surveys in the 1790s. Although Menzies had not gone ashore on the Columbia River, and never traveled inland, he showed Douglas specimens he had collected along the Northwest coast and advised him to look for certain plants indigenous to the region. Botanist Robert Brown, the keeper of the Banks Herbarium where Menzies' collections were stored, added his own suggestions. Aylmer Lambert, who had written a definitive study of pines (a group that at the time included all conifers), encouraged Douglas to collect cones and needles. Dr. John Richardson, preparing for a second Arctic expedition with Franklin, shared his first-hand experiences of inland travel, and he and Douglas made plans to rendezvous in the Canadian Prairies two years hence.

Back in the Society's library, Douglas read the journals of Lewis and Clark and pored over Frederick Pursh's two-volume *Flora Americae*, which described over a hundred and thirty plants gathered by the captains during their expedition. John Lindley, the Society's assistant secretary and a skilled taxonomist, was on hand with advice about plant identification as well as the latest methods of preserving seeds and specimens. And Douglas's childhood friend William Booth, who once had doubted that his friend would ever venture far from home, was now working at Chiswick as a clerk and helped with many details of planning and packing.

An Ocean Covered with Foam

At the end of July, David and his brother John caught a coach from Piccadilly to Gravesend, where the Hudson's Bay Company ship was preparing to sail. Noting that "John was affected at parting with me," Douglas boarded the *William and*

Ann and saw his name entered into the roster as the sole passenger. He would be sharing quarters with the ship's surgeon,
who happened to be a close acquaintance from Glasgow: John
Scouler, a nineteen-year-old medical student whom Douglas
had met in the classroom of William Jackson Hooker. A
devoted naturalist, Scouler had accepted the surgeon's job "as
an opportunity of prosecuting his favorite pursuit," and could
hardly contain his excitement at the opportunity to explore
new lands with an old friend.

The *William and Ann*, commanded by Captain Henry
Hanwell, weighed anchor early in the afternoon of July 27.
By 7:30 that evening, she was stuck hard on a bar called the
"Shivering Sand," where she remained until the next morning's high tide set her back afloat. During a two-week passage down the coast of Portugal and Morocco, Douglas wrote
steadily—recording the water temperature and latitude each
day, trying to identify every passing ship and bird. On August
3, he remarked that it had been exactly one year since he
cruised into New York harbor.

When the ship paused at the island of Madeira to take on
water, the two Scotsmen immediately made their way to the
town market. After sampling banana slices fried in butter, they
toured a local vineyard, where they invested in one-sixteenth
of a pipe of fine wine (about eight gallons), sharing the cost of
£7. Then they headed up the island's tallest mountain, where
they collected insects, reptiles, and odd rocks. Douglas characterized Scouler as "a man skilled in several, and devotedly
attached to all, branches of Natural History," and, as Hooker
may have anticipated, the two fed each other's zeal.

Back at sea, the naturalists entertained themselves by dis-
secting a turtle they had captured, then together chafed at the
tedium of blue-water sailing until a freshening breeze off the

Cape Verde Islands filled the air with red-billed tropicbirds. Soon they were counting scores of storm petrels, dubbed "Mother Carey's chickens" by the sailors. They watched shoals of flying fish skim the waves beside them—sometimes landing right on deck. The pair hauled up seaweed to comb for arthropods, and spent nights gazing at hordes of the small jellyfish called blue sailors (*Velella velella*) and pulses of phosphorescent glow in the ship's wake.

On September 10 Douglas marked their approach to latitude 0°: "The god of the seas paid us a visit, and informed us that he would hold a levee the following day." Sailors traditionally initiated first-time crossers of the equator with a ritual that included a crewman dressed as a sea god who dangled initiates over the rail or dunked them in the sea:

> *Saturday, September 11th Heavy rain during most of last night. . . . At ten o'clock this morning Neptune, accompanied with his guard of honour, fulfilled his promise made last night, when all his unqualified sons had an interview with his Majesty. . . . The day was passed with much pleasure.*

Captain Hanwell commented in the ship's log, "Neptune paid us a visit—no work done."

Two weeks later, surrounded by swarms of land birds and exotic butterflies, the *William and Ann* closed on the coast of Brazil. Douglas readied himself for work—"made some paper bags for seeds against reaching Rio Janeiro"—and when the captain announced that they would remain in port for several days to rearrange cargo, neither of the Scotsmen uttered a complaint. "The approach to Rio is particularly grand," Douglas wrote. "The ground is mountainous, but not rugged, and covered with wood to the summit." Once ashore, he called on several Englishmen living in the cosmopolitan city.

He wandered the marketplaces, watching his captain pay half a dollar for a cabbage and marveling at the riot of luxuriant growth among species that barely survived in English hothouses. He attended a midnight service in a local church and met travel writer Maria Graham, "a lady of much information, and tolerably conversant in botany," who described plants from Chile and the island of Juan Fernandez. He visited the home of a Liverpool expatriate who had built an exotic aviary surrounded by an astonishing array of seventy epiphytic orchids.

John Scouler sought out different company in Rio, beginning with an old friend of Dr. Hooker who supplied him with duplicates from his reptile collections. Scouler was able to converse with many of the foreigners in fluent French, an accepted skill for any educated man of the time that Douglas, as he had learned in lower Canada, sadly lacked. The surgeon spent time puzzling over the rocks of the region, returning after one sweltering afternoon on the hillsides with "my pockets filled with the granite of Rio; my hat outside & inside was pinned full of insects & both my arms full of plants."

Douglas also filled his arms with plants, but experience made him more circumspect about his efforts. He fretted that the incessant rain and moldering humidity would ruin his finds, and complained that for lack of a proper text, he couldn't identify them anyway. In an attempt to get the most fragile cuttings clear of the miasma, he carefully crated two boxes to be placed on the next ship bound for England, then worried that they would arrive in the middle of the Northern Hemisphere's winter. With the *William and Ann* preparing to sail, he carted nearly two hundred species on board to dry at sea. "Just as I stepped in the boat," he wrote, "it began to rain heavily, with thunder and lightning. I had to take off my coat and vest to keep my specimens dry."

As the ship moved south again, more curiosities came aboard. Off the mouth of the Rio Plato, immense shoals of seaweed covered the surface. Captain Hanwell, responding to the entreaties of his naturalists, took soundings and drew up long skeins for Douglas and Scouler to examine. They measured one that was sixty feet in length and found starfish and a variety of shells tangled among its strands. "The roots of this plant were a treasure to the zoologist, & might be called a menagerie of marine animals," Scouler observed.

Below the Tropic of Capricorn, pelagic birds descended on the ship in great numbers, including beautiful pintado petrels, fulmars, and the object of one of Douglas's mythical quests—the wandering albatross. "When sitting on the water the wings are gently raised like the swan; and when rising from the water to soar in the atmosphere they partly walk the water, tipping the surface with the points of the wings for the distance of several hundred yards ere they can raise themselves sufficiently high to soar, which they do with the greatest gracefulness," he rhapsodized.

Although the petrels devoured all the oily chum that the naturalists could throw out but avoided capture, the albatrosses were not so wary. As the ship cut a long arc around Cape Horn through violent squalls of sleet and snow, Douglas baited hooks with pork fat and reeled in forty-nine sooty albatrosses and twenty of the wandering and yellow-nosed varieties. The roughest of weather brought the greatest success: "It is only when the wind blows furiously and the ocean is covered with foam like a washing-tub that I could take the Albatross." While Douglas fleshed out study skins, Scouler dissected the esophagus, stomach, and internal organs of one of the magnificent birds. Not surprisingly, he found that evening's supper fishy and disagreeable, whereas the sailors smacked their lips, much preferring albatross over their usual fare of salt beef.

Three weeks after rounding Cape Horn and entering the Pacific, (which Douglas deemed "truly pacific"), the lookout sighted Juan Fernandez Island off the coast of Chile. "The island was approached with equal interest by every one on the vessel . . . from the romance connected with its history," wrote Scouler, referring to the marooning that had inspired the story of Robinson Crusoe. The captain dropped anchor in a sheltered bay, and Douglas and Scouler climbed into a dory with a crew sent to collect fresh water.

> As our boat approached the shore, we were not a little surprised to see smoke issuing from a small straw-thatched hut. . . . But our astonishment was still more increased when, on the eve of landing, a person sprang from the thicket behind, saluted us in English and directed our boat to a sheltered creek.

Their unexpected host turned out to be an English drifter named William Clark, who was tending camp for a group of Chilean seal hunters. Upon sighting the approach of the *William and Ann*, he had taken cover in a thicket, fearing pirates. But when he heard Douglas and Scouler conversing in English, he "sprang from his place of retreat, and no language can convey the pleasure he seemed to feel." The two naturalists reveled in Clark's adventurous story and resourceful adaptations to island life; both remarked on his single cooking pot, his log sofa, and his library of seventeen books—among them a fine copy of *Robinson Crusoe*, "who himself is the latest and most complete edition," Douglas quipped.

During their daily forays on shore, Clark led his guests to Cruz Bay, purported to have been Crusoe's residence. They toured an old Spanish fort where abandoned figs, strawberries, and fruit trees were just ripening and gathered a quantity of pears and quince for the ship's cook to make a pudding. Douglas ventured into the mountainous interior, collecting ferns for Hooker and stopping every few minutes to pick the sharp seeds of a ubiquitous grass from his stockings. In a shady ravine, he stretched out on a carpet of delicate groundcover for a nap. Before departing, the naturalists gave Clark a supply of tea and a few articles of English attire, and sowed a quantity of garden seeds "to add to the comfort of a second edition of Robinson Crusoe, should one appear." Douglas, reluctant to leave, described the island in a letter to his friend William Booth as "one of the most enchanting places that can be imagined—so strong were my feelings in its favour that I would gladly stay alone for two or three years."

The ship was twelve days out of Juan Fernandez when its crew celebrated the turn of the new year of 1825 with quarts of spruce beer. A week later, a sooty tern alighted on a deck spar, presaging their sighting of the Galapagos. Reports from a handful of previous visitors had not prepared either Douglas or Scouler for the stark basalt landscape and bizarre assemblage of species that inhabited this cluster of islands. Anchored off James Island, Captain Hanwell obliged their feverish desire to get ashore: "Sent the cutter & 4 men to get Turtle Mr. Douglas & surgeon in the boat." From one brief session, they returned with "about 2 doz. teal killed by Mr. Douglass, six large green turtles, & two land tortoises, & plenty of iguanas." Douglas reported that the tortoise tasted like veal, and the iguana "makes a very good soup, and contains a large portion of gelatin. I relished it very well, but

I do not know how it would do in London or New York."
During two more landings, Douglas strode about the bare
rocks, brushing birds from his hat and gun barrel, well aware
that "everything, indeed the most trifling particle, becomes
of interest in England." He collected forty-five birds repre-
senting nineteen genera, and prepared the majority as study
skins. He pressed 175 plants, most of them new to science.

Then twelve straight days of warm, mold-inducing rain
spoiled most of the plants and all the bird skins save for a single blue-footed booby. Douglas was mortified that he would
return with so little from such a significant spot, and later
recalled his dismay at having to dump his rotting iguanas
overboard. John Scouler hoped that the meager samples they
did save might excite some interest: "The Gallipagos as will
be seen by this very incomplete notice of their productions
are peculiarly rich in the objects of scientific research." Days
after the islands dropped from sight, boobies were still landing in the rigging of the *William and Ann* as if to invite them
back, but another decade would pass before Charles Darwin
arrived in the *Beagle* to take a fuller measure of the place.

The *William and Ann* plowed north for four weeks through
almost continuous rough weather. With the journey clearly
wearing on them, neither Douglas nor Scouler filled more
than a couple of journal pages between the Galapagos and the
Pacific Northwest. Most of Douglas's words during those eighty
days described his relentless capture of albatrosses. He hooked
another of the wave-gliders on February 25 but "was prevented
from skinning this one by the violent storm (for as I have mentioned before, I never could take them but when the water was
in the most agitated state), during which the second mate fell
on the deck and fractured his right thigh." Douglas winced with
empathy for the unfortunate boatswain: "The excruciating

pain which this poor man suffered until the termination of our voyage can hardly be expressed."

As ship's surgeon, John Scouler was responsible for treating the victim; "the fracture was in the middle third of the femur & the upper part of the bone had nearly protruded through the skin." Scouler—who could not have seen too many compound leg fractures in his nineteen years— administered regular drafts of opium and worried that "the quick motions of the vessel and the stormy weather we have renders the uniting of the bone precarious and subjects this unfortunate man to great pain."

The *William and Ann* approached the Northwest coast in March 1825, and finally sighted the landmark of Cape Disappointment on April 3. Douglas recorded that an easterly breeze brought them within four miles of the maw of the Columbia, but "a violent storm from the west obliged us again to put to sea." Two days later, riding a keen northwest wind, Captain Hanwell lined up his ship for another attempt. "Such an opportunity was not lost, all sail was set, joy and expectation was on every countenance, all glad to make themselves useful." Douglas and Scouler helped mark the depth soundings as they eased over the river's dreaded bar, marveling at the steep promontory of Cape Disappointment and the size of the conifers visible on shore.

By four in the afternoon the *William and Ann* had dropped anchor in Baker Bay on the estuary's north side. Captain Hanwell touched off cannon shots to announce their arrival to the traders at Fort George, the fur post situated on the south side of the river near the modern town of Astoria, Oregon. The *William and Ann* received no resounding answer, but the naturalists didn't care. They had been at sea, by Douglas's count, for eight months and fourteen days, and "the joy of viewing land, the hope of in a few days ranging through the long wished-for spot and the pleasure of again resuming my wonted employment may readily be calculated. . . . With truth I may count this one of the happy moments of my life."

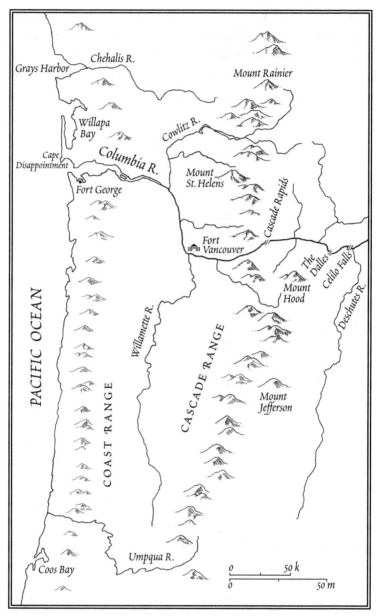

The lower Columbia River and the Northwest Coast, from Grays Harbor to the Umpqua country, 1825–33

III.
BETWEEN THE DESERT AND THE SEA
SPRING–SUMMER 1825

River's Mouth

Safely anchored in the Columbia's vast estuary, the crew of the *William and Ann* "spent the evening with great mirth," and that night Douglas enjoyed the luxury of a sound sleep, undisturbed by the noise and motion of a ship at sea for the first time in nearly nine months. But his anticipation of exploring the new green universe met with disappointment as torrents of cold rain fell without interruption during the entire day of April 8. The next morning, a thick fog prevented Captain Hanwell from proceeding upriver to the Hudson's Bay Company post, but did not deter several tribal canoeists from paddling across Baker Bay in their cedar dugouts, which ranged from torpedo-shaped two-man craft to long vessels sporting elegantly carved prows. Uncertain of their intentions, the captain had the boarding netting put up and stationed sentries about the deck, but he was soon convinced of their friendliness and allowed a small group to climb aboard. Both Douglas and

Scouler commented on the civility of the Chinook visitors, who brought dried salmon, fresh sturgeon, and dried berries of various kinds, "and soon showed themselves to be dexterous people at bargaining for trinkets, molasses, and bread." One of the Chinook men wore European clothes, and his wife knew several English and French words. Scouler and Douglas communicated by means of a few English words and a great many signs, all the while noting with great curiosity the Chinooks' compressed foreheads and the ornaments of shell, copper, and beads that decorated their pierced ears and noses.

Later that afternoon, the two naturalists were rowed a short distance to the shore of Baker Bay. As they stepped out of the boat, their eyes immediately fell upon the leathery green leaves and bell-shaped pink blossoms of salal. Dr. Menzies had advised them to be on the alert for this evergreen shrub, which he believed would be a valuable addition to English gardens. "So pleased was I," wrote Douglas, "that I could scarcely see anything but it." According to Scouler, they soon pried their attention away to "penetrate into those primeval forests never before explored by the curiosity of the botanist," and returned to shore laden with salmonberry branches and early wildflowers. Back aboard the ship, they found that a messenger from Fort George had brought fresh butter and potatoes, which were greeted with relish when the cook served them as part of their evening tea.

The two naturalists again went ashore on April 10, this time crossing the promontory of Cape Disappointment. Dense tangles of the salal they had so admired the day before dogged their every footstep, but upon finally reaching the rocky headland overlooking the Pacific, their efforts were rewarded by the spectacle of an eagle devouring a duck. The next day the weather cleared sufficiently for Captain

Hanwell to maneuver the *William and Ann* across the river
to Fort George, where the crew was welcomed by Hudson's
Bay Company clerk Alexander McKenzie (no relation to the
famous Canadian explorer). Although McKenzie was busy
overseeing the transfer of operations to a new headquarters
some ninety miles upstream, he took time from his duties to
orient his visitors to their new environs. The clerk was married to a daughter of Chief Comcomly, an influential Chinook
headman, and, after serving for five years at Fort George, he
was well acquainted with the geography and culture of the
lower Columbia. When Douglas and Scouler inquired about
signs of unrest among the local tribes, McKenzie explained
that Comcomly had recently lost two of his sons to illness,
despite the ministrations of a neighboring chief with a reputation for healing powers. A third son, believing that his
brothers had actually been murdered by the healer's enchant-
ments, assassinated the medicine chief. McKenzie predicted
that there would soon be more bloodshed between factions
devoted to each headman, and two days later Scouler and
Douglas watched four dozen Chinook men, wearing arrow-
proof elk-hide tunics and wielding scallop-shell rattles, perform a war dance on the beach near the fort.

That same afternoon the Scotsmen met Comcomly, who
was about sixty years old, blind in one eye, and dressed in his
oldest clothing to signify his state of deep mourning. McKenzie
cautioned his guests to address the chief as "Madsu," for one of
his dead sons had been named for him, and it would be considered insensitive to utter the name of the deceased.

On Douglas and Scouler's frequent excursions ashore, they
would have heard the workers and Indians about the fort speaking Chinook jargon, the lingua franca of the lower Columbia.
Most modern linguists believe that the original form of this

jargon, a flexible trade language that borrowed elements from several regional tongues, predated European contact. The tribes began incorporating English words from sailors on ships that came seeking sea otter pelts in the 1790s, then added French terms from fur traders who descended the Columbia two decades later. Douglas was soon tuning his own ear to the nuances of this jargon, remarking that the local pronunciation of the omnipresent evergreen shrub was not *shallon*, as previously recorded by Lewis and Clark, but *salal*.

Undeterred by either tribal disputes or the rain that fell almost daily, Douglas and Scouler explored the countryside at every opportunity. With his pressing paper still buried in the ship's hold, Douglas could not lay in any plants, but he did begin an annotated list of the local flora. He included information regarding habitat, timing of the seed cycle, comparisons to other members of the same genus, tribal usage, and aesthetic judgments. Sometimes he added reminders to himself to be sure and obtain fruits or seeds. In the coming months, he would constantly update this rough tally, keyed to the dried specimens he accumulated as he built the framework for a systematic ordering of Northwest flora.

Most of the first twenty items were common coastal species such as salal, currant, and foamflower. He found a strange vine, "a stately and beautiful shrub," that he thought must be allied to the honeysuckles. He paid close attention to edible plants, watching Chinook children gather young horsetail stems and eat them raw or boiled. Upon sampling one, Scouler pronounced the taste similar to English asparagus.

After their botanical kits were offloaded on April 16, the two naturalists spent three days on the Columbia's south shore, then moved upstream to fill their vasculums on Tongue Point. "We have now so completely ransacked the neighborhood of Ft. George," Scouler remarked, "that very few new plants now attract our notice . . . in this delemna I set out in quest of animals and was tolerably successful." He began to pile up fossil shells from the eroding bluffs and busied himself writing a treatise on coastal geology. Douglas waded through the "dense and gloomy" forest, gazing up at cedars, hemlocks, and firs of intimidating size while lamenting the absence of the deciduous hardwoods that had brought him so much pleasure on the Eastern seaboard. The most common of the towering conifers he thought must be *Pinus taxifolia*, whose branches Menzies had collected along the coast. Named by Lambert for the resemblance of its needles to those of a yew, it would eventually be removed from the pine family and reclassified as a "false hemlock," *Pseudotsuga menziesii*, commonly called the Douglas fir. The collector would have been flattered, for in his opinion it formed "one of the most striking and truly graceful objects of Nature."

On those soggy April days, in the shadows of the spreading fir branches, the common coastal salamanders called rough-skinned newts would have been completing their breeding cycle—just as they had for uncountable generations into the past, and just as they still do today. Any curious walker shuffling through the shiny salal understory is bound to uncover chocolate-colored backs moving at a slow, deliberate pace. Sometimes a mistaken step will goad a newt into a threat posture that reveals its bright citrus underside, a sudden splash of yellow-orange in an otherwise green world. Yet neither Douglas nor Scouler mentioned any newts in their coastal writings. The kneeling naturalists, although keenly sensitive toward this new ground, did not come close to absorbing it all.

The Heart of the District

Rain was still falling on April 16 when John McLoughlin, the head of the Columbia District, appeared at Fort George from the new post upriver. Forty-one years old that year, the six-foot-four chief factor presented an imposing figure. A medical doctor and acute businessman, he had played a key role in the amalgamation of the North West Company with the rival Hudson's Bay Company in 1821. Now he faced the formidable task of coordinating the trade in the vast territory between the Rockies and the Pacific.

From their headquarters in London, the Bay Company's governing committee issued detailed directives to their operatives, and Captain Hanwell presented McLoughlin with a packet containing the annual orders for the Columbia District. Item 11 explained the presence of his unusual visitor.

> Mr. Douglas is a Passenger in the William & Ann, and is sent by the Horticultural Society for the express purpose of collecting Plants and other subjects of natural history. He will remain with you until next season, and we desire you will afford every assistance in promoting the objects of his Mission.

Well aware that he could accomplish little without the support of the fur company, Douglas felt gratified by the chief factor's reaction: "In the most frank and handsome manner he assured me that everything in his power would be done to promote the views of the Society."

After inspecting the William and Ann's cargo, McLoughlin invited Douglas to accompany him upstream to the new headquarters. Leaving John Scouler to his medical duties with the ship, the two men stepped into a "small" canoe manned by a Canadian and six Indians. With the paddlers fortifying themselves on wild raspberry shoots, they made forty miles against a strong current, passing wide bottoms covered with willows and rushes. In camp that evening, Douglas enjoyed sturgeon for supper and slept in the beached canoe. The next morning, after a 3 a.m. start, the naturalist studied the landmark volcanoes named St. Helens and Hood. Small tribal camps dotted the river, with fishermen busily netting salmon and spearing sturgeon.

Along the way, Douglas probably heard from McLoughlin about the continuing uncertainty over the international boundary in the Pacific Northwest. In order to uphold British claims, the Hudson's Bay Company was working to expand their trade and cement amicable relationships with the region's tribes. Although the Americans did not have any trading posts, or indeed any presence at all on the lower Columbia at that time, their ships were trading along the coast, and there was no doubt about their desire to secure as much of the territory as possible during upcoming treaty negotiations. Recognizing that Fort George, on the south side of the Columbia, had been ceded to the United States after the War of 1812, the Bay Company had decided to relocate their depot to the river's north bank. The first suitable building site they could find lay over ninety miles inland, a few miles above the mouth of the Willamette River on a wide alluvial plain known as Belle Vue Point.

It was 10 p.m. before the canoe hauled ashore and the men climbed a steep trail to the newly christened Fort Vancouver, where McLoughlin showed his guest to a pitched tent. When he awoke next morning, Douglas found himself atop a high bluff about a mile from the Columbia, with "sublimely grand" views of snow-capped Mounts Hood and Jefferson beyond the river. Outside the wide gates of the stockade stretched an extensive natural meadow fringed by riparian alders and maples, backed by an endless evergreen forest.

Workers had ploughed a large plot for a garden and were busy planting a hundred barrels of seed potatoes and three acres of peas in the rich soil. Horses and cattle grazed in a pasture near the river, and a pigsty was under construction to house several hogs imported from Hawaii.

Inside the stockade, two storehouses, a trading hall, and temporary quarters for the staff bordered a large court-yard. Carpenters worked to complete a dwelling house for McLoughlin, his mixed-blood wife Marguerite, their children, and his son from a previous liaison in eastern Canada. As he toured the post, Douglas encountered two clerks, both native Scots and both bearing the surname McDonald. Finan, the elder of the two, had helped inaugurate the Columbia beaver trade under North West Company agent David Thompson in 1807, and had worked in every corner of the region.

The younger McDonald, Archibald, (no relation to Finan), had come to the Columbia only three years before to oversee the accounting at Fort George. There he had married Chief Comcomly's youngest daughter, known as Princess Raven. Only a year later, shortly after the birth of their son Ranald, Raven had died, and another of Comcomly's daughters had agreed to care for the child temporarily. The gregarious Archie, who was now stationed at Fort Vancouver, would have wel-comed the arrival of a compatriot with news of his homeland, especially one who shared his interest in natural history, and over time he and Douglas became particular friends.

Three days later, the small society at the fort was augmented by yet another Scot when John McLeod swept downriver in command of the annual canoe brigade from the four interior posts, delivering their past year's harvest of furs. Amidst the ensuing commotion, Douglas could study the business and social hierarchy of the trade. Chief Factor McLoughlin occupied the pinnacle of the organizational chart. Chief Traders represented the middle tier, while numerous clerks completed the ranks of the "gentlemen" who oversaw the trade at the various posts. Some of the fur agents were accompanied by tribal or mixed-blood wives with whom they lived in a form of common-law marriage known as "the custom of the country" in recognition of their distance from clergy or civil officialdom. Many of their children worked in the trade as well, as laborers and sometimes as clerks.

The labor force, referred to as "the men" or "the servants," included French Canadian voyageurs; descendants of early French settlers and eastern tribes, especially Cree; Iroquois Indians who had migrated west with the fur trade; and Hawaiians who had been recruited by ships' captains to work on the Columbia. In all, there were approximately one hundred men on the payroll of the Columbia District. Except for scattered "freemen" who worked as independent hunters for the fur company, and an occasional small party of American trappers who ventured into the Snake River country, these Bay Company men represented the entire non-native population of the Northwest.

As Douglas toured the neighborhood over the next two weeks, he was especially taken with the beautiful sky-blue camas lilies that covered the lowlands along the river. He watched tribal women use pointed digging sticks to unearth great quantities of the nutritive corms, which they called *quamash*. They baked these camas roots in earth ovens, and he found that the finished product tasted much like a baked pear. "Captain Lewis observes that when eaten in a large quantity they occasion bowel complaints," Douglas commented. "This I am not aware of, but assuredly they produce flatulence: when in the Indian hut I was almost blown out by strength of wind."

Along the river and on the outskirts of the forest, he judged two species of lupine, one purple and one white, to be the "most magnificent herbaceous plants I have ever beheld." He squeezed the roots of two fine sunflowers, which exuded a gum that smelled like turpentine; the seeds of one kind were ground into a sort of bread by tribal women. He noted that the natives fashioned salmon hooks from the tough wood of vine maples, but the voyageurs hated walking through thickets of the shrub that they called *bois du diable*. Durable Pacific dogwood served well for canoe masts and spars, but there was another maple he had not yet identified that also satisfied those special applications. Although clearly fascinated by these local plant usages, the collector always remained focused on possibilities for the Chiswick gardens—after watching local people gather the roots of a handsome yellow fritillary called yellow bells, he preserved a selection of the bulbs in a jar with dry sand.

Some of the plants seemed strangely familiar. One lily resembled the asphodel he knew as a child. The tribes' yew bows, he realized, came from the same genus that Englishmen had used for their longbows of yore. A few of the species proved to be recent immigrants. Lamb's quarters grew outside many tribal lodges, and it confounded him that people would ignore this age-old European food to suck on peeled raspberry shoots. Timothy grass would soon be common around the grounds of Fort Vancouver, and several other introductions that he identified, including lance-leafed plantain, spurge, and bentgrass, remain well-known weeds today.

Adapting to travel by canoe and on horseback in place of the carriages and steamboats to which he was accustomed, the collector and his notebook soon became a familiar sight up and down the river. The local Indians he hired as guides and paddlers were fascinated when he lit his pipe by focusing the sun with the magnifying lens he used to study plants. After watching him stir an effervescent powder into a cup of water with his finger, then swallow what appeared to be a violently boiling mixture, they dubbed him "*Olla Piska,* which in the Chenook tongue signifies *fire.*" But his eyeglasses impressed them most: "To place a pair of spectacles on the nose is beyond all their comprehension: they immediately place their hand tight on the mouth, a gesture of dread or astonishment."

Douglas found the local tribes friendly and tolerant of his inquiries. On May 1, during a brief excursion downstream, he passed several tribal sweat lodges "formed of sticks, mud, and turfs, with a small hole for means of entering." He learned that these "steaming huts or vapour baths" were used by hunting and war parties, and watched naked men enter the hut, then exit a bit later to plunge into a nearby pool. But when invited to join them in a sweat, he demurred: "My curiosity was not so strong as to regale myself with a bath."

In early May, John Scouler nipped upstream in one of the canoes ferrying cargo from the ship. Douglas immediately took him on a collecting foray to a river island directly across from the post, named for Archibald Menzies. As they plodded along the shore, they reminisced about Glasgow and a favorite hike to the summit of Ben Lomond in the Highlands. Douglas reported to Hooker that Scouler frequently asked "what would the Dr. give to be with us?" While exploring the island, the friends found a fine forget-me-not, and inspired by the allusion, named the new species *hookeri* in memory of their professor. Fish were also on their minds, so the pair bartered for some whitefish and carefully dissected them. When the surgeon had to return to Fort George, Douglas accompanied him downstream, and at the traditional half-way encampment near the mouth of the Cowlitz River they traded for some suckers to examine. These bottom feeders, unfortunately, were already putrid by the time they reached Fort George. When Scouler boarded the *William and Ann* to cruise north to Nootka Sound, Douglas returned upriver with the remainder of his baggage from London.

Touching the Falls

The extra gear Douglas brought from the coast overfilled his tent, so he moved his quarters to a roomier deerskin lodge. In addition to clothing and bedding, his trunks held vasculums, multiple plant presses, heavy cartridge paper for drying specimens, writing vellum, pens, an inkstand, extra journal notebooks, plant identification manuals, extracts from the journals of Lewis and Clark, a taxidermy kit, a thermometer, a gun, buckshot of various sizes, and a tea kettle. Bundles of pressed plants, skins of birds and mammals, rock samples, and insect specimens completed his décor.

As he resumed his perambulations, he missed Scouler's companionship, writing to Hooker that "we were always great friends, and here our friendship increased. . . . I looked upon myself as very lonely the first few weeks after he sailed." But the collector soon found a distraction when a Cayuse Indian who lived upriver and worked as a courier for the traders visited Fort Vancouver, and Douglas secured his services as a guide for a brief excursion inland. Nicknamed "Yes" by the voyageurs, the Cayuse led the collector on a three-day journey to the rapids known as the Cascades of the Columbia (now drowned below the site of Bonneville Dam). Along the way, he found a beautiful tall white bog orchid. He gathered a biscuitroot whose young stems were eaten by the locals and branches of the shrub called kinnikinnick, whose leaves were dried over a fire and smoked. Competing with elk for an attractive clover that had caught his eye, he was able to collect his perfect specimen "only under brushwood where they cannot go." He found that he was rapidly adjusting to his new mode of outdoor living, comparing attitudes in England, where "people shudder at the idea of sleeping with a window open" to the nonchalance with which he spread his blanket beneath a pine tree. He confessed that he had at first looked on the practice "with a sort of dread. Now I am well accustomed to it, so much so that comfort seems superfluity."

Back at Fort Vancouver, he checked on a small garden plot he had scratched out earlier in the spring and sowed with seeds he had brought from London. Among a bed of turnips, he had transplanted two evening primroses with the hope that they might produce seed and prove "a valuable addition to this handsome tribe of plants." He was pleased to find that both primroses were in flower, two feet high, and branching vigorously. Attuned to the delicate timing of regeneration, he spent the next three weeks in the vicinity of the post, "procuring seeds of early flowering plants and collecting various objects of natural history." He separated each species of seed into a small paper packet, carefully labeled it, then added the letter "S" beside the name of its parent on his master list.

By June 20 all the cargo from the *William and Ann* had been transported upstream to Fort Vancouver, trade goods for the coming year had been sorted and packed for the interior posts, and John McLeod was preparing to lead his brigade back upriver. Several men from the fort were going along in an extra canoe on the first leg of the journey to help with portages, and Douglas took advantage of this opportunity to visit new territory, climbing aboard one of the large fur trade bateaux—outsized canoes that measured up to thirty-four feet in length, seated ten or more paddlers, and hauled over a ton of cargo. With good birch bark in short supply west of the Rockies, the Columbia bateaux were sheathed with split or sawn cedar planks, but remained light enough for the voyageurs to carry around unnavigable rapids.

The first of the laborious portages occurred at the Cascades, and while the men lugged goods bundled into ninety-pound "pieces" around the falls, Douglas paced off the rills of the cataract. Leaning close to bedded layers of limestone and sandstone, he touched petrified trees and collected mineral samples for Scouler. The salmon runs had begun, and Douglas watched tribal families gather at traditional fishing grounds. The collector, who tended to see the world through the lens of plants, was also an avid fisherman. He pored over the tribes' maple-rimmed hoop nets and examined the difference between the willow and cedar barks that served for heavy cordage. He studied their long seine nets woven from Indian hemp, remarking that they used carved cedar floats for corks along a net's top line and oblong grooved stones as weights to sink its bottom strand.

He measured and weighed a couple of thirty-five-pound salmon, then pulled out a supply of trade tobacco he had brought along for just such an occasion. Grown in Virginia, Brazil, or the Caribbean, the tobacco leaves were spun into ropes called "twist" that could be parceled out an inch or two at a time as a gift or in exchange for favors or food. Negotiating the purchase of his fish, Douglas felt he made an excellent bargain: "Both were purchased for 2 inches of tobacco, value two pence, or one penny each. How little the value from that in England, where the same quantity would cost £3 or £4, and not crisped salmon as I have it, cooked under the shade of a lordly pine or rocky dell." Personal comfort often figured prominently in his writings, and that evening, after supping pleasurably on his crisped salmon, he enjoyed a sound sleep on a bed of pine branches.

Over the next four days, the voyageurs fought gusty afternoon winds as they paddled upstream from the Cascades. Douglas watched the woods grow sparser as the climate changed from wet to dry. When the brigade arrived at the long series of rapids known as the Dalles, the east end of the Columbia Gorge struck the naturalist with the same sudden shock felt by all coastal visitors: "Nothing but extensive plains and barren hills, with the greater part of the herbage scorched and dead by the intense heat."

Realizing that the portage around the Dalles would require some time, Douglas decided to walk for a while. Always excited by odd life histories, he studied a broomrape that survived by being "parasitic on the roots of various grasses which have been burned by the natives in the autumn for the purpose of affording a tender herbage in spring for their horses." He watched native women pull thread from a particular sedge, while others twisted strings from cattail leaves to bind large rushes into durable mats. Douglas did not name any of the tribal groups at the east end of the Gorge, which is not surprising, for the area around the Dalles forms a rough boundary between Chinookan and Sahaptin language families, and between Coastal and Plateau cultural patterns. Seasonal visitors from near and far created a cultural and linguistic mélange that would have been impossible for a newcomer to codify.

Between the Dalles and Celilo Falls, his journal entries encapsulated the great paradox of shrub-steppe flowers. "As far as the eye can stretch is one dreary waste of barren soil thinly clothed with herbage," he wrote, and yet his plant list from the area included some of his most successful collections. Antelope bitterbrush and several kinds of sage appeared stately as well as fragrant. He pocketed seeds of the deeply lobed pink clarkia, dedicated to the Corps of Discovery's captain, "an exceedingly beautiful plant. I hope it may grown in England." Douglas found a new phlox that he named after Joseph Sabine, and tagged a penstemon to fellow naturalist John Richardson, who was then employed "on his second hazardous journey to the polar sea." Lupines and mariposa lilies were in full flower along with the giant blazing-star, an annual that thrives among the basalt escarpments of the shrub-steppe.

In Douglas's time, some of the most prominent of the rock faces between the Dalles and Celilo showed carefully pecked patterns of petroglyphs and the streaked red slashes of pictographic art. Their styles reflected the astonishing mix of cultures that made their way to the rapids during the salmon runs to participate in the Columbia's largest trade market. Douglas saw the plants and the people and the fish, but made no mention of any stone faces staring at him while he worked.

By the time he met the canoes upstream at Celilo Falls, his thermometer read 97 degrees in the shade, and his feet "were in one blister," but he had found some very interesting novelties that he was sure would amuse the plant lovers at home. After this long portage was completed, the brigade faced a clear channel for some distance ahead. The "convoy men" who had come to help carry goods turned back toward Fort Vancouver, tucking Douglas's blistered feet securely in the canoe for a quick ride downstream. With specimens from seventy-three species of plants to sort, he concluded "my time I consider well spent."

IV.
TAKING THE SMOKE
SUMMER–WINTER 1825

Cockqua's Hat

After returning to Fort Vancouver, Douglas spent the first two weeks of July sorting seeds and tending to his new finds. In order to prevent mold, he regularly opened his presses and replaced the absorbent blotters that separated the thinner papers folded around each delicate specimen. Retightening the straps of the press, he hoped for warm dry weather. Near the post, he happened upon an abundance of the beautiful creamy phlox he had named for Sabine and devoted himself to amassing a quantity of its tiny seeds. "I hope it will ere long decorate the garden at Chiswick," he wrote.

Gathering seeds was not the only painstaking work at hand, for he also had to analyze each new plant he found, determining its genus and then comparing its characteristics to the limited species that had been identified in the botanical manuals of his day. His primary references were Thomas Nuttall's *Genera of North American Plants* and Frederick

Pursh's *Flora Americae*. During his frequent perusals of Pursh's compendium, he had noticed an appeal to botanists traveling along the Columbia to search for an odd sedge with a potato-like root, used by the Indians for bread, of which Meriwether Lewis had preserved only a single imperfect sample. Aware that the botanical world would welcome a complete example of this mysterious tuber, Douglas hired a Canadian guide and two Indian paddlers to take him downstream in search. His guide, fluent in Chinook, learned that the plant might be found along the south shore near the river's mouth, but to Douglas's irritation, the ongoing tribal dispute made it unsafe to land there.

He decided to investigate the north side instead, where he slogged through heavy rain across Cape Disappointment, accompanied by his small retinue. "Now I have a little idea of traveling without the luxuries of life," he wrote. Even though he carried several oilcloths, his gear was soaked, and he spent an hour each night drying his blanket over a fire before lying down to sleep. But, he added philosophically, "I managed to live very well" on tea and biscuits from his knapsack and fresh fish he caught in creeks along the way.

Upon reaching the ocean, he watched hosts of brown pelicans cruise above the breakers, reminding him of similar birds from the Galapagos. Small gulls on the beach looked like the ring-billed variety he had seen recently at the Dalles. More intriguingly, unfamiliar petrels and an albatross that normally should have been cruising off the continental shelf were plainly visible from the headlands—perhaps, as sometimes happens today, the recent storm had blown them inshore.

He found a few soggy roots that seemed to match the description of Lewis's "bulbous rush," but no stems or leaves to identify the plant. There were consolations, however.

The salal that grew so luxuriantly along the lower river was setting berries, which Douglas pronounced "good, indeed by far the best in the country," and he preserved samples in a bottle of spirits.

During their perambulations, the travelers called on Cockqua, a coastal Chinook chief who had developed a fondness for British manners from visiting sea captains. "After saluting me with 'clachouie,' their word for 'friend,' or 'how are you?' and a shake of his hand, water was brought immediately for me to wash, and a fire kindled." Cockqua led Douglas to a large dugout canoe that contained a fresh sturgeon and invited him to choose a cut for his supper. Before selecting the head and shoulder portion, Douglas measured the beast at ten feet in length and three feet in girth, and estimated its weight at between four and five hundred pounds. The Chinooks followed this meal with a war dance in anticipation of a raid by the Clatsops, and Cockqua invited Douglas to sleep in his lodge for safety's sake. But the collector, declaring that "fear should never be shown," declined the offer.

Next morning, the villagers treated their visitor to a spirited test of marksmanship, in which several men showed off their abilities. Douglas took up the challenge, shooting a bald eagle on the wing, then blasting the crown from a hat that his counterpart tossed in the air. "Great value was then laid on my gun and high offers made," he bragged. "My fame was sounded through the camp." Since European and American ships had been anchoring in Baker Bay for more than three decades, Douglas certainly was not the first white man to show off his musketry skills to the tribe; he may have inflated the people's response to his showmanship, or the villagers may have been applauding to be polite.

While camped near the village, Douglas watched the Chinook women gather and prepare a variety of edible plants. They pried the large tendrils of a beach peavine from the sand to eat raw, and roasted those of a seashore lupine. They cooked tasty sword fern roots over embers, and strung the green fronds as decorative garlands. Inside Cockqua's lodge, Douglas examined "baskets, hats made after their own fashion, cups and pouches, of very fine workmanship." He noticed the colors and patterns created by an astonishing variety of fibers whose particular qualities the weavers had put to optimal use: leaves of cattail and bear-grass; roots and the twisted inner bark from cedar trees; a strange brown seaweed; and a variety of sedges.

When Douglas prepared to leave, he repaid the villagers for their hospitality with gifts of tobacco, knives, nails, and gunflints. Cockqua presented his departing guest with a set of gambling sticks fashioned from carved spirea shoots tipped with beaver incisors, one of his own hats, and "a promise that the maker (a little girl twelve years of age, a relation of his own) would make me some hats like the chiefs hats from England." Douglas ordered four, including one with his initials woven into the crown.

Back at his Fort Vancouver headquarters, he updated his master list, now approaching the mid–four hundreds. He made several day trips in the vicinity, checking on the state of ripening seeds among the plants that interested him most. With stacks of dried specimens overwhelming his deerskin lodge, he once again sought more commodious accommodations, this time in a hut sheathed with cedar bark. The rains continued to fall, and he complained that the dampness was "greatly retarding" his efforts. Actually, the area of Scotland where he grew up receives slightly more annual precipitation than the vicinity of Fort Vancouver, and the Highlands, where he botanized with William Hooker, are at least as wet as the lower Columbia. But his garden was doing well, and both of his evening primroses were starting to set seed.

Smoking Together

By mid-August the collector was ready for another jaunt, and on the 19th he joined a hunting party headed up the Willamette River. Leading the group was the tall, redheaded Finan McDonald, who had a reputation for both pugnacity and kindness. Finan had sought pelts on both sides of the Oregon Cascades, and paddling up the smooth waters of the lower Willamette, he better than anyone could have explained to his guest the toll taken on the region by the fur trade. Though trappers had been working the Willamette for only a little more than a decade, Douglas learned that "this at one time was looked on as the finest place for hunting west of the Rocky Mountains. The beaver is now scarce."

Thirty-five miles upstream, their progress was interrupted by Willamette Falls. Although there were more than two dozen men in the party, the collector's account sounds as if he were lugging the boat all by himself: "I had considerable difficulty in making the portages at the Falls, having to haul the canoe up with rope; this laborious undertaking occupied three hours, and one hour on my return."

Continuing south through the valley, Douglas thought the rich alluvial soil held great possibilities for cultivation. About twenty miles above the falls, McDonald made camp near a Kalapuya tribal village, and his hunters soon brought in seventeen black-tailed deer. Douglas was very interested in this coastal species and took careful measurements of both sexes, but was able to preserve only one of the animals.

He had better luck with smaller mammals, preparing skins of a dormouse, a pocket gopher that would later bear his name, and a "small rodent" (probably a ground squirrel). He developed an affection for the band-tailed pigeon, a beautiful plump relative of the passenger pigeons he had seen in New York. These birds gathered in the evening around a salt spring near the camp, where their "elegant movements when picking up and licking the saline particles that were found round the edge afforded me great amusement." He easily shot five of them, but again was able to keep only a single "miserable" skin. Although Douglas consistently touted his marksmanship, he usually denigrated his skills as a taxidermist.

The collector's ten-day jaunt up the Willamette added more than two dozen species to his master list, including a small purple iris later designated with *douglasii* for its species name. Two other plants of great interest came from the Kalapuya village. The first was a native tobacco, for which he had been on the alert since his arrival. Up until this point in his stay, he had seen only a single tobacco plant, in the hand of an Indian at the Cascades who, to Douglas's great frustration, refused to exchange it for any consideration—not even for two full ounces of English tobacco. He had been searching diligently ever since, and when he came across a hidden garden in a small forest opening near the village, he "supplied myself with seeds and specimens without delay."

Douglas, however, was much less alone in the wild than he imagined, and the owner of the plot soon caught up with him. Avoiding what could have been a serious affront—tobacco had a spiritual significance in many tribal rituals—Douglas whipped out his own manufactured strain as a peace offering and called upon his Chinook jargon to inquire about growing methods. The Kalapuya man accepted the gift, then described the tribal technique of searching out an open area in the woods with plenty of downed trees. They burned the deadfall, then planted tobacco seed in the ashes. Douglas, long a fan of "the good effects produced on vegetation by the use of carbon," was well pleased. "When we smoked," he concluded at the end of his account for plant #447, "we were all in all."

His second discovery involved a new conifer. Douglas noticed some unusually large pine seeds in the tobacco pouch of one of the Kalapuya men, who said that his people chewed the tasty nuts, which came from the mountains beyond the Willamette, in the Umpqua country. Finan McDonald and his party were headed in that direction, and the collector offered a handsome reward to anyone who would bring back cones and viable seeds of the fabulous pine.

Since Finan and his hunters were planning to spend the winter exploring the area to the south, Douglas returned with two men in a small canoe to Fort Vancouver at the end of August, "fraught with the treasures I had collected." He quickly cataloged his new findings and departed again—fall was in the air, and there were seeds to gather. At the beginning of September, accompanied by a Canadian from the post, the naturalist moved upriver with a local headman called Chumtalia who lived near the Cascades of the Columbia. Douglas claimed to be the first white man to journey to the portage without an armed escort, but under his guide's "very attentive" protection, the bold adventurer experienced no difficulties at all.

After setting up his tent near Chumtalia's village, Douglas asked the voyageur to drench the ground with water to stave off fleas, which seemed to be particularly bothersome around fishing sites. The next day, leaving most of his gear in the Canadian's care, he followed Chumtalia's brother into the steep mountains north of the river. For any naturalist seeking new habitats and a view of the surrounding country, the obvious landmark to aim for is Table Mountain. Clearly visible from the river, Table's raw triangular face still hints at the massive landslide that formed the Cascades of the Columbia five centuries ago.

Douglas and Chumtalia's brother climbed about two-thirds of the way up their chosen peak before making camp the first night. Expecting to reach the summit the next day and return to camp before nightfall, Douglas decided to travel light, leaving behind his blanket and all of his food except three ounces of tea, a pound of sugar, and four small biscuits. But rugged terrain, intersected by streams and strewn with downed trees, slowed their progress, and night was falling by the time they reached the top, where "all the herbage is low scrub and chiefly herb plants." He managed to bring down a bald eagle for supper, which they roasted and found to be very good eating. He brewed a pot of tea, "the monarch of all food after fatiguing journeys," and served it in cups fashioned from tree bark. Then he and his guide spent a frigid night with their feet near the fire.

The late-season plant communities on the loose, rocky slopes of Table Mountain did not offer the collector much that was new, and the return trek left him so footsore that he had to recuperate for a couple of days at Chumtalia's village. To pass the time, he amused himself "with fishing and shooting seals, which were sporting in vast numbers in the rapid where the salmon are particularly abundant."

When sufficiently refreshed, he and Chumtalia made an excursion into the mountains on the south side of the river. "The ascent was easier than the former one, and I [that is, he and Chumtalia] reached the top after a labourious walk of fifteen hours." The pair may have headed up Eagle Creek and into modern Oregon's Mark Hatfield Wilderness Area. This stream, or any of several smaller drainages along the Columbia upstream from Multnomah Falls, plunges from the foothills of Mount Hood through a lush, diverse forest. Although the fire history of the Columbia Gorge shows that a blaze of cataclysmic proportion swept the area around 1700, Douglas and Chumtalia would have wandered through pockets of untouched old growth that increased as they climbed. Above two thousand feet, Douglas would have encountered his first Pacific silver fir, then passed through huge stands of western hemlock and Douglas fir. When the pair reached an altitude around four thousand feet, they would have noticed the dancing branches and purple-black plates of noble fir, widely acknowledged as the stateliest member of its family.

In order to scientifically define these two new species, Douglas needed cones and seeds. Since mature firs bear most of their cones in the very crowns, he struggled to find a solution. He tried shooting down a cone-laden branch, but his buckshot didn't reach the target. He considered felling a tree, but his two-pound ax was of little use against such massive trunks. He climbed one likely candidate, only to find that the crown wouldn't bear his weight, possibly because he insisted on wearing his knapsack: thanks to his unfortunate experience in Oak #43 on Lake St. Clair, he had "learned the propriety of leaving no property at the bottom of a tree." Back on the ground, he wrote himself a memo to obtain seeds "by some means or other."

He also met with frustration when he tried to obtain samples of the bear-grass that grew in the high country. He could not find one of the thick basal clumps in either fresh enough flower or ripe enough seed to suit his exacting taste. But he was able to confirm Frederick Pursh's assertion that some of the Columbia tribes wove watertight baskets from bear-grass leaves, and he did learn that the local name for the plant was *Quip Quip*.

In both Coastal and Plateau basketry traditions, the dark purple-brown shoots of bear-grass are often imbricated with lighter material to create distinct geometric patterns. Douglas might have seen this effect when "my Indian friend Cockqua arrived here from his tribe on the coast, and brought me three of the hats made in the English fashion, which I ordered when there in July." Cockqua explained that the fourth hat, with Douglas's initials woven into the crown, was not yet finished. The collector paid for the entire set with a seven-shilling blanket, as well as "a few needles, beads, pins, and rings as a present for the little girl" who crafted them. Cockqua also delivered a packet of jet black evergreen huckleberry seeds, dried according to Douglas's instructions. They produced some of the most delicious fruits of their family and were good candidates for cultivation.

A Rusty Nail

Back at Fort Vancouver on September 13, he found that "Mr. Scouler had taken possession of my house." The surgeon had sailed with the *William and Ann* as far north as the Queen Charlotte Islands, doctoring people of all races and ages. The friends regaled each other with accounts of their adventures, "unconscious of time, until the sun from behind the majestic hills warned us that a new day had come." On his way upriver, Scouler had collected a yellow-flowered lily, and he hoped that "this plant might be named in honor of Mr. Douglass, who has been so zealously employed in collecting the vegetable productions of the N.W. Coast."

From Scouler, the collector learned that Captain Hanwell was planning to point the *William and Ann* for England without delay. Since Douglas wanted to send as many of his gleanings home with her as possible, he spent the following week in a frenzy of packing. His future reputation, after all, depended on how well his collections endured the long ocean voyage back to London.

When Captain Hanwell paid a visit to Fort Vancouver in early October, Douglas's packing was almost finished; as far as he was concerned, the situation could not have worked out better. His shipment contained sixteen large bundles of dried plants, accompanied by copious notes. Of the four hundred and ninety-nine species on his list, "I have as far as in my power collected at least twelve of a sort where they could be had or double that number of most," he wrote to Hooker. There were also hundreds of seed packets, bottled bulbs, and twined roots. The bottle of rum-soaked salal berries, however, was not on the packing list, for an Iroquois working at Fort Vancouver had sampled the contents and had found the concoction quite palatable. Douglas added a box of bird and mammal skins and another of Indian artifacts, presumably including the goods and gifts obtained from Cockqua. In case something happened to the *William and Ann*, the naturalist kept a chest containing over a hundred varieties of seed.

After Scouler and Hanwell departed from Fort Vancouver to prepare the ship for sea, Douglas continued his work. He made a copy of his journal for the Horticultural Society, as requested by Joseph Sabine. He penned letters to his brother John and his friend William Booth. He informed Archibald Menzies of the treasures growing on his namesake island. To William Hooker, he wrote an enthusiastic account of his botanical finds, poking a bit of fun at his zealous collecting habits: "In my walks in the Forests I could not suffer a plant to remain in the ground." He inquired fondly after Hooker's family, and worried that the children would not remember him. "I should like to see Master William, is he still fond of Botany? He has long since forgot me."

A letter to Governor DeWitt Clinton summarized all that had happened since they had last seen each other in New York. Filling three pages with dense script, Douglas recounted the rigors and rewards of his voyage around Cape Horn: "Gladly would I spend a few days on the shores of Tierra del Fuego . . . it would be interesting to compare the geography of the plants found on the lowest extremities of the Southern continent with the very extensive & beautiful flora of the northern in similar situations." By the time he turned to the subject of the Columbia, he had used up his paper and had to cross-write the rest of his story in perpendicular lines neatly penned atop his ocean adventures. It would be curious to know whether Governor Clinton ever found time to sit down with his magnifying glass to decipher them.

While loading one of his last boxes at Fort Vancouver, Douglas caught his knee on a rusty nail, and over the next few days the wound grew infected, preventing him from escorting his boxes downstream to see them safely aboard the *William and Ann* as he had intended. He sent a note to Captain Hanwell, "requesting he would have the goodness to place them in an airy situation, particularly the chest of seed, and, if possible, above the level of the water." The captain replied that he would gladly honor the request, and Douglas entrusted his precious cargo to one of the canoes busily transporting fur packs to the ship.

His knee was still bothering him on October 22, when he received word that the *William and Ann* remained at the river's mouth, waiting for a favorable wind and tide to cross the bar. He had been planning another visit to the coast and impulsively decided to set out right away "in a small canoe with four Indians for the purpose of visiting my old ship-mates on my way." The paddlers were beating into a strong headwind when they collided with a deadhead stump that sent them ashore for repairs. While the men worked on their boat, Douglas cooked tea and salmon, then convinced them to paddle through the night. Despite a predawn wind that forced them to portage across Tongue Point, they made Fort George by 9 a.m. There, members of Comcomly's band informed Douglas that the *William and Ann* had set sail an hour before. "This was a severe disappointment," he wrote.

While his exhausted paddlers napped, the disappointed collector roamed the point looking for ripe seeds. He returned at dusk to call at Comcomly's lodge, still respectfully calling him Madsu. He met the chief's brother, Tha-mux-i, who was about to return to his home near Grays Harbor, exactly where Douglas wanted to go. Comcomly sent for a large dugout canoe to ferry them across the river, and with a dozen paddlers they made the passage despite a fierce wind. A shaken Douglas attributed their survival only to "the strength of the boat and dexterity of the Indians." His flour and tea were ruined, but a few ounces of chocolate and some rum survived, and when they landed safely on the north shore, he offered watered-down drams all around, even though sharing alcohol with the tribes was strictly forbidden by McLoughlin. Tha-mux-i, whom the white traders had nicknamed The Beard, flatly refused, explaining that he once had caused such a row after drinking that his neighbors were forced to tie him up, and he hadn't touched a drop since. He did accept a gift of tobacco, and astonished Douglas by his deep long gulps on the pipe. When Tha-mux-i saw how timidly the naturalist puffed, he asked, "Why do you throw away the food? See, I take it in my belly."

Douglas started north on October 26, along with Tha-mux-i and four of his relatives, who carried their canoe across the narrow peninsula that divides the Columbia from Willapa Bay. The rain fell in torrents as they started across the shallow bay, and by the time they cleared Ledbetter Point, the wind was "producing an agitation on the shoal water frightful in the extreme." Abandoning the canoe, they continued overland to Grays Harbor. Their fifteen-mile journey passed through a landscape of sloughs and wetlands, and they were soaked by the time they reached a fishing village where Tha-mux-i had planned to purchase food. But the place was deserted, and they resorted to kinnikinnick berries for their fare.

After reaching the south shore of Grays Harbor, the party bivouacked in a stupor, munching on the roots of wapato and seashore lupine, "a very nutritious wholesome food called in the chenook language *Somuchtan*." Douglas made himself "a small booth of pine branches, grass, and a few old mats," but his blanket was so drenched that he spent the night by the fire, unable to sleep, and "the following day found me so broken down with fatigue and starvation, and my knee so much worse, that I could not stir out."

By the next morning he was able to stir far enough to bring down five ducks with one lucky shot from his musket. Such a blast might seem like one of the collector's exaggerations, but fall migration still covers Grays Harbor with dense rafts of waterfowl and skeins of hungry shorebirds. His gunfire alerted Tha-mux-i's village on the far side of the harbor, and they dispatched a canoe to pick up the travelers. Restored by roast duck and a cup of tea, the naturalist was soon shooting at gulls and band-tailed pigeons on the bay, and gathering bear-grass and a sedge used for cordage on shore.

Tha-mux-i's village was one of several located around the mouth of the Chehalis River, near the modern town of Aberdeen. Alexander McKenzie, the clerk from Fort George, was trading in the area, and he and Douglas decided to return home by ascending the Chehalis, crossing a divide to the Cowlitz River, then following that stream to the Columbia. Tha-mux-i obligingly ferried the white men sixty miles upstream in his canoe. Citing "the zeal and kindness I had received at his hands," Douglas agreed to the chief's request for a proper English shave.

At the upper Chehalis village where they disembarked, McKenzie and the naturalist traded twenty rounds of ammunition, two feet of tobacco, several gunflints, and a packet of vermillion for a pack horse and guide to lead them to the Cowlitz. Complaining about bad trails and high prices, they wound around the base of Mount St. Helens to the village of Schachanaway, who supplied them with nourishment and a canoe for their downstream trip to the Columbia. Hoisting Douglas's cloak and blanket as sails, the travelers reached Fort Vancouver at midnight on November 15.

Douglas was far from satisfied with this twenty-five-day journey. Most of the berries he collected for seed stock had been consumed, and McKenzie had "suffered severely from eating the roots of a false asphodel." The collector had left all his birds and seaweeds behind, bringing back only bear-grass and a few currant seeds to show for his troubles. His infected knee remained so swollen that "from this period to the end of December my infirm state of health, and the prevalence of the rainy season entirely precluded any thought of Botany."

First Winter

On the contrary, Douglas's journal indicates that he continued to think about botany a great deal. A fine artemisia from the banks of the Cowlitz and a new evening primrose raised the total of his plant tabulation to five hundred and ten. An unfamiliar monkeyflower reminded him that he hadn't yet collected mature seeds of a particularly handsome variety of that genus, but he intended to soon find some.

There were other interesting developments to consider as well. Three days after Douglas returned to the post, the fall express came in from Hudson Bay, led by Alexander Roderick McLeod (no relation to John McLeod). Unfortunately, the canoes had departed the bay before the annual ship from London arrived there, so McLeod carried no fresh mail, but he did bring several passengers, including a mixed-blood woman whose father ran a post near the Rockies and who would soon become the new wife of widower Archibald McDonald. McLeod had also crossed paths with Douglas's London friend, Dr. John Richardson, the surgeon on John Franklin's latest Arctic expedition, who sent news of their progress.

During the course of the winter, Douglas learned that Alexander McLeod had spent much of his career in the Northwest Territories around the Peace and Mackenzie rivers, becoming proficient in the native languages and gathering knowledge about Arctic geography. The veteran trader told entrancing stories of a Russian fur trade post between the Yukon and Stikine rivers, displayed five- and six-gallon malleable-iron pots of Russian manufacture, and assured his listener that in the tundra "the difficulty of transportation by land or water is trifling."

As Douglas dreamed of the icy north, a more liquid form of precipitation seeped into his cedar bark hut. On

Christmas Day, he finally accepted Dr. McLoughlin's invitation to move his belongings into the chief factor's "half-finished" house. Douglas's bad knee forced him to miss a yuletide horseback outing, and on New Year's Day 1826, he penned a somewhat melodramatic assessment of his position on the planet.

> I am here now, and God only knows where I may be the next. In all probability, if a change does not take place, I will shortly be consigned to the tomb. I can die satisfied with myself. I never have given cause for remonstrance or pain to an individual on earth. I am in my twenty-seventh year.

During the next two months the author of this melancholy passage perked up enough to crawl about in search of mosses, collect numerous birds, and pen lively prose sketches of the fauna he had encountered during his travels in the region. He caught the complexity of the bald eagle's place on the river, how "although powerful, it is overcome by several other species"; how the natives perfectly imitated its weak *chuck-chuck* whistle in their name for it; how it took years of molting through distinct stages to reach the magnificent adult plumage. Below the Cascades he shot a bald eagle that had just snatched a sturgeon from the water, and was amazed to find that he could not release the raptor's talons from the fish without plunging a needle into one of the bird's neck vertebrae.

The naturalist recognized the intimate and evolving relationships between small birds and the river's human inhabitants. He watched Steller's jays gather in groups of several dozen around the kitchen middens of Indian villages. Magpies picked at the scabs on horses' backs in the Fort Vancouver corrals. Both crows and ravens ate carrion and hung around encampments, but to Douglas's surprise, he found that the crows were both less shy and less common.

He apparently obtained much of his information from company hunters, who told him that the gigantic "California buzzards" (California condors) ranged far upstream, following the salmon, and assured him that they favored quills from condor primary feathers for their tobacco pipe stems. The Indians whom Douglas questioned extended the condors' range farther south, intimating that the large vultures might winter in the Umpqua Mountains south of the Willamette.

When he asked about the three species of deer that we now call mule, white-tailed, and Columbia black-tailed, the hunters described the animals' distribution east to the Flathead region of the Columbia Plateau. For the species the voyageurs called *chevreuil*, Douglas recorded a Chinook jargon name, *Mowitch*. Beyond that, he related the female call for its young as "mæ mæ, pronounced shortly." It was this call that Kalapuya tribal hunters imitated when they cut a joint of the large, hollow-stemmed wetland plant known as cow parsnip and blew on it like a flute.

> With this, aided by a head and horns of a full grown buck, which the hunter carries with him as a decoy, and which moves backwards and forwards among the long grass, alternately feigning the voice with the tube, the unsuspecting animal is attracted within a few yards in the hope of finding its partner, when instantly up springs the hunter and plants an arrow in his object.

Douglas took hints about new species from every possible source, including the furs he saw tribal people wearing. That is how he discovered a strange white fox that differed from the common member of the family. "The first that came under my notice were two skins forming a robe for an Indian child . . . I was desirous of purchasing some but too great value was put on them."

Douglas saw his first "lynx" (more likely a bobcat) when he and Alexander McLeod went on a February hunting excursion with a small bull terrier that leaped at the cat's throat and brought it down. In late February, he got another pleasant surprise when Jean Baptiste McKay, one of Finan McDonald's hunters, returned from the Umpqua country and presented him with the prize for which Douglas had offered a reward the previous summer: an extraordinary pine cone that measured sixteen inches in length and ten inches in girth. "Unquestionably this is the most splendid specimen of American vegetation," Douglas reported to Scouler. McKay also brought a wealth of information he had gathered. The tree attained the enormous height of two hundred feet, he reported. The trunk was remarkably straight, the wood of fine quality. With John McLoughlin serving as interpreter—the naturalist's French was still apparently not up to snuff—Douglas supplied McKay with paper bags in which to gather more seeds of the intriguing pine.

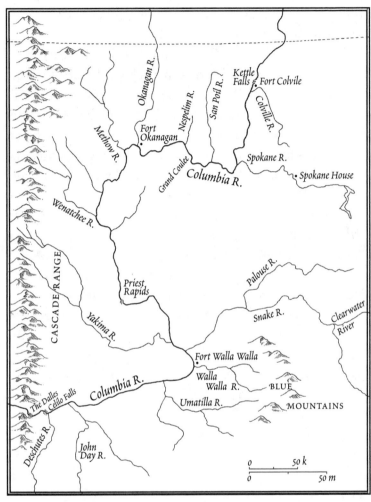

The interior of the Columbia District, from the Blue Mountains to
Kettle Falls, 1826–27

V.
THE INTERIOR YEAR
SPRING 1826

The Costume of the Country

B y spring 1826, Douglas had resided in the Columbia District for almost a full year. He had witnessed an entire life cycle of the flora of the lower Columbia, had collected thousands of seeds and specimens, and might in good conscience have boarded the next ship home. But he was not ready to leave—the combination of his knee injury and uncooperative weather had prevented him from obtaining many desirable plants. He had not visited the headwaters of the Columbia or probed the foothills of the Rocky Mountains. He had barely touched the arid lands of the Columbia Plateau, which Lewis and Clark had only skirted and no naturalist had yet penetrated.

Assured of the continuing support of John McLoughlin for transportation and lodging, Douglas decided to disregard his original instructions from Joseph Sabine, hoping it "may not incur his displeasure." In a journal extract sent to the Horticultural Society, he justified his decision.

> *From what I have seen in the country, and what I have enabled to*
> *do, there is still much to be done; after a careful consideration as to*
> *the propriety of remaining for a season longer than instructed to do,*
> *I have resolved not to leave for another year to come. . . . I cannot*
> *in justice to the Society's interest do otherwise. . . . If the motive*
> *which induces me to make this arrangement should not be approved*
> *of, I beg it may at least be pardoned. Most cheerfully will I labour*
> *for this year without any remuneration, if I get only wherewith to*
> *purchase a little clothing.*

The impending departure of the spring express, a pair of
light, fast-moving canoes that carried dispatches from the
interior posts to Hudson Bay, offered an opportunity to travel
inland. On Monday, March 20, Douglas left Fort Vancouver
with two boats heading upstream under the command of John
McLeod, the trader with whom he had traveled to Celilo
Falls the previous summer. Dr. McLoughlin sent along a note
explaining the collector's mission to the agents at the interior
posts, and allotted room in one of the canoes for over a hundred pounds of precious collecting paper, "an enormous indulgence." Douglas accordingly skimped on his other baggage,
limiting his wardrobe to one extra shirt, two handkerchiefs,
a blanket, a single cloak, and no stockings at all. "Thus," he
explained, "I adapt my costume to that of the country."

With three pigs and three calves swelling the passenger
list, the express moved along the now-familiar stretch of river
to the Cascades of the Columbia, where Douglas plucked several new mosses from the noisy edges of waterfalls that were
running hard with snowmelt. While the voyageurs portaged
the rapids, he admired rainbows that formed above the roiling
water. Continuing through the Gorge, he listened to the melancholy echo produced by waves pounding the walls of rock

caverns and collected blossoms from the same currant bushes that had provided him with ripe fruit the previous fall.

On Thursday evening the party camped in a small cove at the base of the Dalles. Douglas had visited this place twice before without incident, but on this occasion about four hundred and fifty men gathered around the encampment. Disgruntled by the skimpy amount of tobacco that McLeod presented, their "manners announced anything but amicable feelings towards us." The next morning, Douglas used his knowledge of trade jargon to play the role of mediator. "Finding two of the principal men who understood the Chenook tongue, with which I am partially acquainted, the little I had I found on this occasion very useful." Additional gifts of tobacco mollified tempers for a while, but farther upriver at the portage around Celio Falls, tribal men began splashing water on the fur traders' flintlocks and refusing to allow their bateaux back into the river. When McLeod attempted to push one of the offenders out of the way, another man threatened him with a bow and a handful of arrows. Douglas "instantly slipped the cover off my gun, which at the time was charged with buckshot, and presented it at him, and invited him to fire his arrow."

A tall Cayuse suddenly stepped into the breach and admonished the ruffians, defusing the situation within minutes. This Plateau headman and three of his young men accompanied the two fur trade canoes a few miles upriver to camp, then backtracked to the portage to make sure that no more mischief was in the making. McLeod rewarded their rescuer generously, and Douglas added his own token of gratitude.

> I being King George's Chief, or the "Grass Man," as I am called,
> bored a hole through the only shilling which I possessed, and which

*had been in my pocket ever since I left London, and observing that
the septum of his nose was perforated, I suspended the coin to it by
a bit of brass wire, a ceremony which afterwards proved a seal of
lasting friendship between us.*

While helping stand guard through the night, Douglas lit
one of his precious candles and scratched out a letter to Dr.
Hooker. He justified his bold decision to stay in the Columbia
country for another year, and wished that his teacher could
have been present many times to help sort out the Northwest
mosses. He relayed McLeod's news of the Franklin Expedition
and their hopes of marking out a Northwest Passage, and
confessed interest in learning more about the Russians, who
apparently wore long beards and were very wicked. With sleet
beginning to pelt his tent, Douglas's thoughts turned homeward. "I hope you, Mrs. Hooker, and all the family are in good
health. I should much like to see William to have his old story
of the *Poor Scotch Caterpillar*, which he so often told me."

The collector spent much of the next three days walking
on shore to keep warm as the canoe party traced the river's
sharp bend north through Wallula Gap. The open plains were
sprinkled with sage and antelope bitterbrush "and other shrubs
which to me were perfectly unknown and the whole herbage
very different indeed from the vegetation on the coast." Soon
after passing the mouth of the Walla Walla River on Tuesday
afternoon, the brigade landed at Fort Walla Walla (also called
Fort Nez Perce), where Douglas "was received with much
kindness by Mr. S. Black, the person in charge."

Samuel Black was a fur trade veteran with a checkered
background who had worked for both the North West and
Hudson's Bay companies, mostly in the northern Prairie
districts. Only recently posted to Walla Walla, he had seen

little of the territory beyond his trade house. When Douglas mentioned his desire to explore the Blue Mountains east of the post, however, he found the agent more than willing to help. After judging the most auspicious month for a plant expedition, Douglas made plans to return early in June from his exploration upriver.

John McLeod's party lingered only two days at Fort Walla Walla before proceeding upstream on the Columbia. They stopped for breakfast at the mouth of the Snake River, where Douglas collected early lupines and primroses that were starting to peek out. On April 1 they passed the mouth of the Yakima River and entered the windswept Hanford Reach. This was disorienting country for someone accustomed to trees, and Douglas saw the famed White Bluffs as "mountains of white clay, with scarcely a vestige of herbage or verdure to be seen."

While the crew lugged gear and livestock past Priest Rapids, the collector, using his plant press as a lap desk, wrote a letter to John Scouler in which he confided that his eyesight had been deteriorating over the winter: "Without pain or inflammation, a dimness has come on which is a great loss to me, especially with the use of the gun which, as you know, I could handle to some advantage." Douglas repeated his plans for further exploration and looked forward to rehashing their adventures in some of their old Scottish haunts when he returned. Then, perhaps goaded by a call from McLeod at the top of the portage, he hurriedly closed his letter with a collector's tease: "Excuse this bad writing. I have little time, and less convenience for writing . . . this is penned upon the top of my specimen board, under which are some exceedingly interesting things."

The Whole Chain of the River

It had been a heavy snow year, and as the paddlers labored around the Columbia's Big Bend, the ground remained blanketed in white. Douglas saw occasional bears, wolves, foxes, and badgers scurrying through the dryland shrubs, and he kept an eye out for the dancing leks of sharp-tailed and sage grouse. When the canoes reached the fur post at the mouth of the Okanogan River, lingering snow stymied any botanizing, and Douglas was relieved to move on. While the voyageurs fought through Nespelem Canyon, he spotted some bare ground and tramped into the sagebrush, enjoying the bright yellow lichens that festooned the oldest twisted trunks.

At the mouth of the Spokane River, they met a party of fourteen men under John Dease, the "commandant in the interior" who would host the naturalist from his base at the new Fort Colvile. Dease and his men planned to camp on the Spokane for a week or more, which was fine with the collector—the snow had receded, exposing "an extensive plain, with groups of pine trees, like an English lawn, and rising bluffs or little eminences clothed with small brushwood and rugged rocks sprinkled with Ferns, Mosses, and Lichens." The cryptogrammic crust was exploding with spring wildflowers, and he thought "this part of the Columbia is by far the most beautiful and varied I have yet seen."

Here Douglas parted ways with the express crew, who were continuing on to Hudson Bay. He entrusted McLeod with a note for Governor George Simpson, the company's chief of operations in North America, asking that an enclosed packet of seeds and mail be placed on the first possible boat to England. Not forgetting his manners, the collector assured the governor that he was being well cared for along the course of his travels.

*With infinite gratification do I mention the friendly attentions and
assistance I have experienced from every person connected with the
Company in particular John McLouglin Esqr. I find this Country
so exceedingly interesting that I have resolved to devote the whole
of this Season in the Upper Country.*

Before sealing his letters, Douglas added a postscript to
William Hooker, extolling the virtues of several new phlox
species and ending with the classic line of a naturalist in the
April shrub-steppe: "I can hardly sit down to write, not knowing what to gather first."

Camped at the mouth of the Spokane River, the collector dried paper that had been splashed during the trip upriver
and began a new specimen tally. His first three dozen items
included four different desert parsleys or biscuitroots (members of the genus *Lomatium*), important food plants of Plateau
tribes. He couldn't distinguish the species—botanists today
still struggle to classify this genus—but did sniff the strong
caraway odor of one variety. After pressing buttercups and
shooting stars, whose angular petals "impart a grace to the
scanty verdure of American spring," he made forays in every
direction, "more for the purpose of viewing the soil and face
of the country, with any bird or animal I might pick up."

There was plenty of human traffic around the campsite,
including chief agent John Work, a familiar face from the previous summer at Fort Vancouver. Work had been raised on an
Irish farm, and although he had poor handwriting, a plodding
approach to business, terrible French, and little formal education, he displayed a keen understanding of the natural world.
Work had already been collecting mammals and seeds and
sending them to Fort Vancouver. One of his packets had
allowed Douglas to identify "Wormwood of the Voyageurs"

as antelope bitterbrush; another alerted the naturalist to the fine hawthorns of the interior. As they talked about the flora and fauna of the upper Columbia, Work described the bighorn sheep that inhabited the highlands, called *mouton gris* by the voyageurs. Douglas had admired a mounted bighorn in Peale's Museum in Philadelphia, and longed to collect a set of horns and a pelt to send home to England.

After assigning men to drive the calves upriver, John Dease's party loaded the three pigs into a canoe and departed the Spokane campsite on April 19. During the ninety-mile paddle to Kettle Falls, Douglas watched the landforms change from basalt plains to rugged granite and limestone mountains, while the vegetation thickened from open shrub-steppe to mixed coniferous forest. He identified lodgepole and ponderosa pines as well as Engelmann spruce, but was most impressed by the western tamaracks, measuring several up to thirty feet in circumference and pacing off one fallen tree at 144 feet to the first branch, straight and clear. On the second day the voyageurs portaged a substantial rapid that, surprisingly, had never been named. Douglas, showing his awareness of the explorers who had preceded him, suggested they call it Thompson Rapids after David Thompson, "the first [white] person to descend the whole chain of the river from its source to the ocean."

Heavy rain poured down as the party splashed ashore near Kettle Falls after three days on the river. With the new Fort Colvile still in the early stages of construction, they pitched camp on "a small circular plain where the new establishment is to be." Once the tents were set up, they dined in style on steelhead trout and dried buffalo meat; after the meal Douglas removed his specimens from their double oilcloth wrap and set about the familiar process of changing blotting papers to prevent mold, then arranged his plant presses beneath pieces of bark near the fire to help them dry.

For untold generations, Kettle Falls had served as the most important fishery and trade center for the Salish-speaking tribes on the middle and upper Columbia. Although the peak of spring runoff and the tribal gatherings centered around summer salmon runs remained months away, Douglas waxed poetic over the setting.

The whole stream is precipitated over a perpendicular ledge, twenty-four feet high, besides several smaller cascades, which shiver the water into the most picturesque snowy flakes and foam for the distance of one hundred and fifty yards, where a small oval rocky island, studded with a few shrubs and trees, separates the channel in two.

Spring showers of rain and snow dominated the next two and a half weeks, hampering his efforts to do much more than keep his plants dry. In spells of fair weather he circled the post, following the snow line as it ascended the hills. He pawed through carpets of yellow glacier lilies, resolving to return for mature seeds, and tested the "low moist peaty soils" where trillium bloomed in abundance. He examined the rusty scales on the underside of buffalo berry leaves, and nibbled the corms of spring beauty (Indian potato), which tasted distinctively different than ones he had sampled on the coast. When he wasn't digging and snipping plants, Douglas, bad eyesight or no, was firing away at grouse and long-billed curlews. The North American variety of this curve-billed shorebird differed from the European one, and he had the luck to stumble on a nest with one of their beautiful eggs, "a light brown with blue spots, the small end more pointed."

During all of his shooting, Douglas damaged his musket, and no one at the nascent Fort Colvile had the skill to fix it. Dease recommended the services of his former clerk, Jaco Finlay, "being the only person within the space of eight hundred miles who could do it." Finlay lived a few days' journey south, but two of his sons happened to be visiting Fort Colvile and agreed to escort the naturalist to their father.

Jaco's Clan

For the journey, Dease contributed a horse for Douglas to ride and another to carry his blanket and supplies for botanizing along the way. On May 9, as they traveled south through the Colville Valley, high water forced them to skirt flooded meadows and cut across one low mountain. In the higher elevations, Douglas shot seven dusky blue grouse, counting thirteen eggs in the ovary of one female. He also stumbled on a grouse nest with seven newly laid eggs, effectively camouflaged by "bright brownish-dun with large and small red spots." The practical naturalist blew one of the eggs for his collection, then scrambled its contents to complete "a comfortable supper" of dried buffalo and fresh grouse.

Near noon of the second day, they reached a ford of the Little Pend Oreille River (near modern Arden, Washington) to find the rivulet too swollen for the horses to wade. Faced with either building a raft or swimming, the men reached a unanimous decision: "Being all well accustomed to the water, we chose the latter." Floating on his back, Douglas hefted musket and paper over his head and frog-kicked across as a sudden hailstorm pelted the river. He made a second trip for his blanket and extra clothes while the Finlay boys ferried the saddles on their heads. He did not mention how he transported the hollow grouse egg.

Early the third morning, as the men crested the divide between the Colville and Spokane drainages, the woods opened up to offer "one of the most sublime views that could possibly be, of rugged mountains, deep valleys, and mountain rills." The switch from mixed coniferous forest to the drier, more open habitats of bunchgrass and ponderosa pine appealed to Douglas. The tribes managed this landscape with periodic ground fires, so the orange-barked pines were mature and well spaced. The trail wound down to the Spokane River (near present-day Tumtum), then followed grassy benches upstream to its confluence with the Little Spokane. There, in the summer of 1810, Jaco Finlay, working as an emissary of David Thompson, had constructed the fur post known as Spokane House. When the Hudson's Bay Company relocated to Fort Colvile, they had bequeathed the buildings to Jaco.

Finlay, a man of mixed Cree and Scottish descent, was married to a Spokane woman named Teshwintichina, and several of their children lived at Spokane House. So did some of Jaco's offspring from previous wives, providing Douglas with a sense of how Cree, French, and Scottish people were woven into the culture of the northern Plateau.

On the pleasant spring day that Douglas rode into Spokane House, the fifty-eight-year-old Finlay received his guest with a meal of camas, bitterroot, and cakes of black tree lichen. The collector had already tasted plenty of camas and some bitterroot, but the inky black lichen cakes were a novelty to him. Although he made no mention of Jaco's wife or daughters, one of these women must have explained the process that Douglas recorded: how they gathered the black lichen that grew on certain trees, then carefully cleared it of twigs before soaking it in cold water. When the lichen was perfectly soft, they arranged it in an earth oven atop heated stones, protecting it below and above with layers of grass and dead leaves. After a full night of baking, they raked away the coals and shaped the sticky goo into cakes while it was still warm.

While Jaco repaired the broken firelock, Douglas wandered upstream on the Spokane River, which would have led him past Nine Mile Falls. Upon his return, he was so delighted to find his firearm restored that he paid Jaco with a full pound of tobacco. That was a great deal of smoke, and may have represented either the importance of his musket or a down payment on interviews concerning the local scene, for Jaco had talents in addition to gunsmithing: "Mr. Finlay being a man of extensive information as to the appearance of the country, animals, and so on, Mr. Dease kindly gave me a note to him requesting that he would show me anything that he deemed curious in the way of plants, &c."

The collector had many questions for the elder Finlay, but communication proved to be an unexpected problem. "Unfortunately he did not speak the English language, and my very partial knowledge of French prevented me from obtaining information which I should have acquired."

Guided by one of the Finlay sons, with whom he apparently communicated just fine, Douglas headed for the hills next morning. With his restored musket he brought down a snipe, always a thrill for a sportsman. He shot ripe cones and dwarf mistletoe clusters from the crown of a ponderosa pine and sampled the variety of shrubs that spotted the open slopes. Ninebark "needed to be saved." Golden currant, with its fine yellow flowers, offered great possibilities. That evening, with the language barrier magically lifted (or with one of his children interpreting), Jaco explained the succession of fruit and ripe seed for the golden currant and two of its relatives in fine detail. Douglas engaged him to collect seeds of any other currants he might run across, as well as corms of a certain onion—"as large as a nut, and particularly mild and well-tasted."

That afternoon one of the boys dragged in a grizzly bear he had shot; while grateful, Douglas had no room to carry such a large pelt back to Fort Colvile. He showed much more interest in Jaco's knowledge of the *mouton gris*. Finlay described these bighorns as extremely shy and promised to keep an eye out for them later that summer, when he and his family gathered huckleberries in the mountains. Jaco may also have provided an interesting bit of information that required no translation: "Their voice is the same as the sheep, *Ma-aa-a*." Douglas was eager to obtain both a sheep and other mammal pelts, and "to Mr. Finlay's sons I offered a small compensation if they would preserve for me the skins of different animals, showing them at the same time how this should be done."

A Middle Spokane village lay a short distance downstream from Spokane House, and during his visit Douglas met the band's chief, Ilum-Spokanee, and some of his family. Between the village and the post, a small copse of trees sheltered a tribal burial ground. "Implements, garments, and gambling articles" surrounded each grave. Near one burial site lay the skin, hooves, and skull of the deceased's favorite horse. Small bundles dangled from the trees, "tied up in the same manner as the provisions which they carry when traveling." Douglas inquired about the significance of the utensils and bundles, but found that "it is difficult to gain any information on these subjects, as nothing seems to hurt the feelings of these people so much as alluding to their departed friends." The naturalist, however curious, seemed to respect this reticence.

On May 13, guided by one of the junior Finlays, Douglas set off on his return trip to Kettle Falls. Their second evening out, they again swam the Little Pend Oreille River, and this time the cold exertion brought on a painful attack of rheumatism in Douglas's shoulders. Unable to sleep, he rolled out of his blanket with first light, brewed some tea, and set off smartly at 4 a.m., walking until he worked up "a profuse perspiration which considerably relieved my suffering."

Back at Fort Colvile he developed a severe headache and fever, and dosed himself with "salts and a few grains of Dover's powder," a patent medicine that contained powdered opium and was used to treat a variety of ills. Afraid of a relapse, he spent the next two rainy days inside his tent, tending to his pressed plants. When the weather cleared, he emerged full of energy, strong enough to follow a most productive routine for a full fortnight. Each day he walked as hard and far as he could in one direction, snapping up every new plant in sight. Nights were devoted to identifying species, sorting seeds, and updating his numbered list.

During Douglas's stay at Fort Colvile, trader William Kittson, a good-natured veteran of most of the posts in the Columbia District, willingly indulged the collector's natural pursuits. On May 22 the two took a canoe trip with a pair of tribal paddlers up the Kettle River, known then as the Dease River or "Sheep Rivulet," which may suggest why Douglas was so eager to explore its upper reaches. While Kittson and the Indians struggled against a current raging with snowmelt, Douglas hopped along the rocky shore, sometimes ascending cliffs that took him high above the river. They didn't see any sheep, but along the way he shot both barn and violet-green swallows; unfortunately, his buckshot tattered the little birds to pieces. He had better luck with a large male dusky blue grouse, which completed two fine pairs of this previously undescribed species.

Around Fort Colvile, Douglas tracked a variety of lilies that bloomed in abundance, including a deep blue wild hyacinth (*Brodeia*) that seemed to like light sandy soils. He developed a taste for a purplish wild onion that flourished on the riverbanks. "This plant is the only vegetable that I have to use in my food; I get it generally stewed down in a little dried buffalo-meat or game." He watched Salish women dig a desert parsley with sulfur-yellow flowers that grew in low swampy areas. The spindly tubers tasted somewhat like parsnips when raw, but were less appealing to his palate when cooked. "The roots are gathered by the natives and boiled or roasted as an article of food (taste insipid). Called by them *Missouii*."

At dawn on May 26, the collector set out on foot to tour the hills south of the post. By noon he had downed a curlew and a sharp-tailed grouse, and feeling stifled by heat that registered 86 degrees on his thermometer, he gave in to the fatigue of his labors and napped in the shade of a cedar tree. It was early evening before he awoke, twenty miles from the fort. By the time he straggled in, a concerned John Dease had arranged for two men to set out in search of him first thing the next morning. Relieved to learn that no accident had befallen his guest, "on my informing him of my delay, he laughed heartily," the sheepish Douglas wrote.

Such misadventures became the stuff of legend among the fur men, and stories about Douglas still persist in the Colville country today. For years a rumor circulated that some of his plants, collected along the placer-rich Columbia upstream from Kettle Falls, were shipped back to England with gold dust glittering in the dirt stuck around their roots.

Yet Douglas, who collected minerals up and down the river's length, never mentioned any gold. Another story credits him with discovering the vein of galena on the shore of Kootenay Lake that was later developed as the Bluebell Mine. But Kootenay Lake lies miles beyond the limit of any of his excursions around Kettle Falls.

One piece of lore, however, does bear the ring of truth. Agent Alexander Anderson, stationed at Fort Colvile in April 1850, wrote a report decrying the scourge of Blackfeet raiders around Flathead Post on the Clark Fork River in Montana: "We have to lament the death of one of our servants, David Finlay, a son of the late David Douglas, who was murdered by a party within 3 miles of the house." In a summary of the year's events, he described David Finlay as an "Interpreter at the Flatheads. He was met by a party of Blackfeet who had just been repulsed from an attempt on the establishment. They shot, stripped, and scalped him."

Anderson's account seems like an odd place to come across the name of David Douglas, more than twenty years after the collector's last visit to Fort Colvile. But since Douglas was so well known around the Northwest—the agent who reported his death was a nephew of Archibald McDonald—it seems unlikely that anyone in the close-knit fur trade community would mistake either the name or the connection. While there is no way to match Douglas with any specific Finlay girl, one of Jaco's daughters would have been about sixteen during the collector's season in the Colville country. Baptized in a Catholic mission as Marie Josephte, she was called Josette.

If the Finlay family did raise a child sired by David Douglas, the taking of the father's first name and the mother's last seems like a satisfactory arrangement. That son would have been about twenty-two years old when the Blackfeet caught up with him three miles shy of Flathead Post in Montana. And if David Finlay did belong to the summer that David Douglas spent in the Colville Valley, it could explain where the collector came by his recipes for lichen cakes and baked *Missouii*.

VI.
SLEEPING ON SHATTERED STONES
SUMMER 1826

Jumping

During the first week of June, the traders at Fort Colvile packed the season's furs to ship downstream to Fort Vancouver. Douglas, who was planning to catch a ride with the brigade as far as Fort Walla Walla, crated his latest treasures and climbed into a canoe with William Kittson. They shoved off from Kettle Falls at 7 a.m. on June 5, and "as soon as our boats got into the current, they darted down the river with the velocity of an arrow just loosed from the bowstring." Kittson reckoned the water to be about sixteen feet higher than when he had paddled upstream a month before. After a mere half hour, they had covered the eight miles to the newly christened Thompson Rapids, which the steersman Pierre L'Etang pronounced "in fine order for jumping"—that is, he meant to run them. Douglas and Kittson, preferring the fine order of solid ground, watched from shore.

No language can convey an idea of the dexterity exhibited by the Canadian boatmen, who pass safely through rapids, whirlpools, and narrow channels . . . where you think the next moment must dash the frail skiff and its burden of human beings to destruction against the steep rocks, these fellows approach and pass with an astonishing coolness and skill, encouraging themselves and one another with a lively and exulting boat-song.

L'Etang delivered the party to the mouth of the Spokane River by midafternoon, having covered a rollicking ninety miles in the space of eight hours. While the men reordered their jumbled cargo, Douglas revisited his former camp, where several new flowers were in bloom. Back on the river, vivid orange-red blossoms of globemallow reminded him of cultivated fields of red poppies from Great Britain. Even after stopping several times for Douglas to botanize, L'Etang's crew made forty more miles to the mouth of the Sanpoil River before stopping for the evening. By early afternoon of their second day, they were coasting into Fort Okanagan, a remarkable performance by experienced voyageurs in fast water.

John Work was in charge at Okanagan that season, and the affable Irishman showed the collector a female sage grouse and a male dusky blue grouse that he had preserved, "both very well done, with a few eggs of the former." Douglas, fascinated with the different species of grouse that inhabited the interior, was collecting information about their habits and habitats at every opportunity. Lacking the time to build a secure box to transport Work's gifts downriver, he arranged to store them at the post until his return.

Less than eighteen hours after beaching at Fort Okanagan, Kittson's canoe was back in the water, now running beside four other bateaux laden with furs from the

Fraser River country to the north. William Connolly, chief factor of the New Caledonia District, greeted Douglas with "genuine and unaffected friendliness . . . and instantly begged that I would consider myself as an old acquaintance." John Work, a font of knowledge about the countryside, manned another of the boats. The collector thoroughly enjoyed the camaraderie of the traders as they sped downriver, but when two of the boats suffered damage in the Rock Island Rapids (below modern Wenatchee), he happily excused himself to explore the rocky shore. With his thermometer registering 92 degrees in the shade, he uncovered an interesting white-blossomed penstemon, along with a small buckwheat, a rare phlox, and a three-foot rattlesnake resting beneath a pile of stones. After a short night punctuated by sheets of dry lightning, Connolly aroused the brigade at first light. They paused at Priest Rapids for a breakfast of fresh salmon and buffalo tongue, then cracked on around the White Bluffs and past the mouth of the Snake River to pull ashore at Fort Walla Walla by nightfall on June 8.

Looking forward to making up some of the sleep he had missed on the way downriver, Douglas paid his respects to Samuel Black, then stretched out on the floor of the Indian hall. When an immense herd of fleas interrupted his antici-pated repose, he moved outdoors into the sagebrush. There his blanket was invaded by "two species of ants, one very black and large . . . the other small and red" (probably car-penter and harvester ants). As soon as it was light enough to see, he gave up all pretense of rest, sharpened a new quill, and completed a letter to Joseph Sabine recounting his recent finds and detailing his future plans. Whether describing the flower and fruit of antelope bitterbrush or the berries and lupines that he knew would thrive in the Chiswick gardens,

Douglas's words conveyed his thrill in the chase. On a more practical level, he explained the difficulties of transporting his collections: "Often I am under the necessity of restricting myself as to the number of specimens, that I may obtain the greater variety of kinds." He closed with a plaintive request for mail: "There is nothing in the world could afford me greater pleasure than hearing from you, and my other friends." After handing his letters to the departing John Work to deliver to Fort Vancouver along with his latest boxes, the collector spent the afternoon transferring information on his recent finds from his pocket notebook to his journal, resolving to go to bed early and get some rest.

Just as he was turning in that evening, his long-delayed sleep was again postponed by an Indian courier, who brought word that the Hudson's Bay Company supply ship had reached the Columbia. The messenger carried a packet from London, and Douglas was struck by the coincidence "that I should write to England in the morning and receive letters on the same day, for in this uninhabited distant land the post calls but seldom." The post had not called for Douglas in almost two years, since his departure from London in July 1824, and he eagerly scanned the handwriting of the senior clerk of the Horticultural Society and his friend William Booth. "Till two hours after midnight I sat poring over these letters as if repeated reading could extract an additional or a different sense from them; and when I did lie down, little as I had slept lately, I never closed my weary eyes," he wrote.

Although gratified by the news from London, Douglas was distressed that he had received no word from his family or from Sabine. A message from John McLoughlin informed him that additional mail was being held at Fort Vancouver and would be forwarded soon. This intelligence left Douglas pining for deeper contact with his motherland. "Never in my life did I feel in such a state of mind. An uneasy, melancholy, and yet pleasing sensation stole over me, accompanied with a passionate longing for the rest of my letters."

Over the next week, he occupied himself with day trips along the Walla Walla River, carefully gleaning the blossoms of the dry sandy plains. He was able to correlate many of these species with Lewis and Clark's descriptions from the same area, but others he was quite certain remained unknown. His new tally sheet steadily approached one hundred and fifty, weighted now toward the various bunchgrasses that thrived in the arid lands. His satisfaction at advancing scientific knowledge was balanced by physical pain, for the beavertail cactus along the river pierced his shoes, and the intense summer sun took a toll on his eyesight.

> *June 15th, Thursday. My eyes began to distress me exceedingly; the sand which blows into them, with the reflection of the sun from the ground, which in many places is quite bare, having made them so sore and inflamed that I can hardly distinguish clearly any object at twelve yards distance.*

The same stormy westerlies that were stirring up sand also halted all fishing on the river, and Samuel Black slaughtered a horse to substitute for the accustomed fresh salmon. Douglas supplemented this "Tartar stew" with roasted "ground rats" (probably Columbia ground squirrels) that he captured in burrows beneath sage and antelope bitterbrush. The local Walla Walla and Cayuse tribes called this rodent *limia*, and he judged the taste "somewhat rancid, or rather of a musky flavor, probably from the bitter strong-scented plants on which it feeds."

The Confines of Eternal Snow

During Douglas's initial visit to Fort Walla Walla three months earlier, he and Samuel Black had discussed the possibility of exploring the Blue Mountains that rose to the southeast. Now, with spring approaching the high country, Black offered provisions and horses, and recommended a Cayuse guide. On the evening of June 16, Douglas sent a message to his escort to be ready the next morning at sunrise.

While finishing a journal entry that evening, he fell sound asleep. Some time later, he awoke to find a "very strange species of rat in the act of depriving me of my inkstand, which . . . was lying close to my pillow." Picking up his gun, "which, with my faithful dog, always is placed under my blanket at my side," he blasted the marauder to eternity. Retrieving the remains, he proceeded to examine the jet black whiskers, enormous ears, long pointed nose, and white belly of a bushy-tailed wood rat. Regretting that the large shot in his musket had spoiled the pack rat's skin, he tamped a lighter load into his barrel and lay in wait. "A female of the same sort venturing to return some hours after, I handed it a smaller shot, which did not destroy the skin." The arrival of daylight revealed that the rat family had "devoured every particle of seed I had collected, eaten clean through a bundle of dried plants, and carried off my soap-brush and razor!"

Douglas's reference to his "faithful dog" marks the first mention in his journals of a canine companion. None of his writings reveal where or when he got his pet, although in later years he would be trailed by a Scottish terrier named Billy. Whether his faithful sleeping companion at Fort Walla Walla was Billy or another pup, he apparently was not a very vigilant mouser.

Short of sleep, Douglas felt his patience wear thin as the morning progressed. His guide didn't appear until 8 a.m., then plodded through certain matters that had to be addressed before they could embark. First, Douglas, speaking English, explained his intentions to Black, who communicated them in French to a Canadian interpreter, who translated the words into the Cayuse language. The guide then insisted on negotiating the fee for his services. After that was settled, he requested a bonus of new shoes, leggings, a knife, some tobacco, and a strip of red cloth for a cap. This bargaining "occupied two hours and was sealed by volumes of smoke from a large stone pipe." Luckily, Douglas did not suffer from a shortage of tobacco to supply the stone pipe, for his new friend William Connolly, shortly before departing for Fort Vancouver, had presented him with the generous gift of twelve feet of twist:

> This article, being, as it were, the currency of the country, and particularly scarce, will enable me to procure guides and to obtain the cheerful performance of many little acts of service, and it is therefore almost invaluable to me.

By midday, three horses were loaded with Douglas's equipment and enough pemmican, dried salmon, biscuits, sugar, and tea for a ten-day journey. As they prepared to depart, Black offered to send along the twelve-year-old son of the fort's interpreter, a lad named Little Wasp who spoke both French and Cayuse. Douglas gratefully accepted, realizing he could use some help communicating with his guide, and the trio set off along the banks of the Walla Walla River. That night they camped on the edge of a small spring in a pleasant birch grove. Noisome mosquitoes roused Douglas at 2 a.m., just in time to sit through a heavy rain shower that, much to his relief, cooled the air and settled the blowing sand.

At The Nose, a distinct escarpment of eroded basalt that looms over the divide between the Walla Walla River's two branches, the guide turned up the south fork, and as they gained elevation the character of the landscape began to change. Long basalt ridges formed a canyon whose south wall supported a rich coniferous forest. On the open rocky slopes of the north side, an abundance of delicious golden currants were just reaching the perfect ripeness for snacking. Entire tracts of the country glowed with the beautiful yellow blossoms of sulfur lupine, reminding the collector "of the 'bonny broom' that enlivens the moors of my native land." Douglas, as was his habit when he approached new ground, pushed the pace; they reached the foothills by 5 p.m. and made camp for the evening. After enjoying a comfortable supper, the collector was so excited by the fresh prospects ahead that "to save time in the morning I made a little additional tea in the evening and left it in the kettle overnight."

They reached timberline early in the afternoon of June 19, having covered over forty miles in two days. The gain in altitude had moved them backward on the seasonal clock, and Douglas noticed a honeysuckle, just budding out, of the same species he had seen in bloom at the mouth of the Spokane River six weeks before. The following day they ascended to the snow line, the border between winter and spring for a botanist. In a sunny exposure on the edge of a large snowfield, he was bending over a purple anemone when clumps of an unfamiliar low plant caught his attention. He fingered its red stem and lobed, alternate leaves, then pried a large, forked root from the ground. It was a most attractive peony, with drooping maroon petals that brightened toward the edges into a fine yellow. Douglas had glimpsed a few "feeble enervated plants" of similar type on the lower slopes, but in the high country the species grew in wonderful abundance, "flowering in perfection on the confines of perpetual snow." Asian peonies were among the most desired cultivars of the Horticultural Society, and Douglas was well aware that he had just come upon the first known member of the peony tribe in North America. After a period of consideration, he attached the name of botanist Robert Brown to his prize find, which has been called Brown's peony ever since. "This valuable addition will I trust be an acquisition to the garden," Douglas confided in his journal. "If in my power, seeds of it must be had."

Seeds, however, were an objective for a future expedition; this was the season for pressing leaves and blossoms. The luxuriant green sward of another new lupine drew him farther upward. By late afternoon, the horses were floundering through deepening snow, and Douglas realized he would have to complete the hike on foot. Leaving Little Wasp and the guide to care for the animals, the collector tucked a few sheaves of paper under his arm and set off alone, with about a thousand feet of elevation to gain before he reached the ridgetop.

The snow was soft at first, but as he climbed, the crust hardened and held firm. Upon reaching the summit, he recorded a temperature of 26 degrees and exulted in conquering such mountainous country, "the snows of which had certainly never been pressed by European foot before." That might have been true, as long as he didn't count French Canadian hunters and mixed-blood trappers as Europeans. He may have been standing on Bald Mountain (just east of Tollgate, Oregon), overlooking the Umatilla and Grande Ronde drainages, where "the view of the surrounding country is extensive and grand."

Douglas had originally intended to descend to the Grande Ronde valley, which he had heard about from some of the Hudson's Bay men. But before he could assess any possible routes, a black cloud descended on the ridge, bringing a fury of thunder, hail, and wind. At times, lightning streaks "would appear in massy sheets, as if the heavens were in a blaze; at others, in vivid zigzag flashes at short intervals with the thunder resounding through the valleys below." Beset by hail, Douglas hurried back to camp, but no fire could withstand the wind and rain. While the storm continued to rage, he grabbed a cold supper, then stripped off his wet clothes, wrapped himself in a wool blanket, and fell into an exhausted sleep. He awoke at midnight, benumbed by cold, but when he tried to stand up and make some tea, he "found my knees refused to do their office." The storm had passed, so he lit a fire and scoured his rheumatic joints with a rough towel to bring them back to life. Then he sipped his tea and glumly considered his situation: "If I have any zeal, for once and the first time it began to cool. Hung my clothes up to dry and lay down and slept until three o'clock."

That would have been 3 a.m., and when day finally dawned, he found the spirits of the Cayuse and Little Wasp equally dampened. Douglas ventured that there must be a low-elevation route that would lead to the Grande Ronde, but his guide protested that such a path would involve swimming the dangerous Umatilla River at high water. Moreover, he wanted no part of a walkabout in territory where Snake (Shoshone) people might be lurking—the Cayuse were currently in dispute with that tribe, and the party would risk losing all their supplies and horses.

Douglas relinquished his hopes and spent the next four days working his way back to Fort Walla Walla at a leisurely pace, gleaning plants "until I had all that appeared peculiar to that district of the hills." He devoted the next two days to putting his new specimens under the press and shaking seeds from plants he had hung to dry before leaving, then immediately began to prepare for another excursion into the mountains. But his guide refused to go along, and quickly departed the scene when Douglas tried to persuade him with "a little corporal chastisement." Apparently Little Wasp had convinced the Cayuse that the white naturalist was a powerful medicine man who could transform anyone who displeased him into a grizzly bear and leave him to run in the woods for the rest of his life. Upon learning the circumstances, Douglas admitted, "it is not to be wondered that these fears acted powerfully on the Indian, and caused him to behave in the way he did."

By midmorning, Samuel Black had located another guide who was less fearful of necromancers. Though the new man had a reputation as a "knave," he did not smoke at all, and Douglas willingly hired him. With a minimum of fuss they set off, aiming this time for the North Fork of the Walla Walla River.

In the high country above that branch, Douglas found giant *frasera*, "a strong plant" as tall as a man, with clusters of green flowers; the exquisite mountain ladyslipper; and a small fragrant white orchid. But he saw little else that was new, and returned in disappointment to the post "much worn down and suffering great pain from violent inflammation of the eyes." During the first week of July, unsettled weather kept him either packing up his plants for fear of rain or spreading them out to dry after a shower passed by.

On July 5, Finan McDonald and Jean Baptiste McKay, Douglas's traveling companions from the previous summer, rode into the fort on their return from a hunting excursion. They exchanged their horses for a boat to carry them to Fort Vancouver, and Douglas seized the opportunity to send along a box containing pressed plants as well as the skins of three ground squirrels and the delicate female pack rat. Then, longing to see the rest of his mail, he impulsively hopped into the canoe alongside his crate.

Afternoon westerlies whistling upriver slowed their progress, and they spent a hungry, mosquito-plagued night. During a stop for breakfast the following morning near the John Day River, a tribal man slipped Douglas's knife from the string that secured it to his trousers. Upon missing his blade, the naturalist offered tobacco for its return; when no one stepped forward, he began patting down the bystanders (the presence of the six-foot-four Finan McDonald might have helped in this process). When he found the knife, he gave the thief such a licking that "I will engage he does not forget the *Man of Grass* in a hurry."

At the Dalles the following evening, they crossed paths with John Work, Archibald McDonald, and William Connolly, ushering the interior brigade with its season's complement of trade goods back upstream. There were additional passengers along, including Finan's Kalispel wife, Peggy, and four of their children. Work delivered Douglas's long-awaited letters, which he "grasped greedily and eagerly broke open." They contained family news from his brother John, instructions from Sabine, and greetings from Dr. McLoughlin and other agents posted downriver. Once again, Douglas's emotions welled over: "I am not ashamed to say . . . I rose from my mat four different times during the night to read my letters; in fact, before morning I might say I had them by heart—my eyes never closed."

At sunrise the naturalist threw his knapsack in the canoe with John Work, who had just lost an ax and a dog while portaging the Cascades and could congratulate Douglas on still having his knife at his side. As they paddled back toward Fort Walla Walla, the collector walked on shore by day and met the canoes each night to camp, happy to share a meal and a pipe with his Columbia acquaintances.

Bubbling Springs

At Fort Walla Walla, the canoe brigade continued up the Columbia while John Work and Archie McDonald traveled east to a large Nez Perce camp to purchase horses for the inland posts. Always eager to see new country, Douglas joined the twenty-eight-man party on July 17 as they canoed up the Snake River through harsh basalt plains. The traders paddled from daybreak till midmorning, then rested in the heat of the day before putting back on the river during late afternoon. Douglas approved of this schedule, measuring temperatures up to 106 degrees and declaring that except for abundant good water, the country was little better than "the burning deserts of Arabia." To relieve himself from the oppressive heat, he took to bathing in the river both morning and evening, "and although it causes weakness in some degree, I have some doubt if I had not I should not have been able to continue the trip."

After six days they reached the confluence of the Snake and Clearwater rivers (now flanked by Clarkston, Washington and Lewiston, Idaho), where they found a gathering of Palus and Nez Perce bands that Douglas numbered at around five hundred men; Work's lesser estimate of two hundred men would still add up to an encampment of over a thousand people. That evening Douglas observed some of the ceremonies of these Plateau tribes, who greeted the traders "with all the pomp and circumstance of singing, dancing, haranguing, and smoking, the whole party being dressed in their best garments."

While the slow and serious work of horse trading proceeded, Douglas enlisted one of Archibald McDonald's men, called Cock de Lard, to accompany him on an excursion to "the spot pointed out to me by the Indians where Lewis and Clarke built their canoes, on their way to the ocean, twenty-one years ago." Then, assured by Work and McDonald that the "horse fair" would occupy a few days, on July 25 the naturalist and "my companion and friend (guide he could not be called, as he was equally a stranger to this country as myself)" headed southeast, toward today's Craig Mountain. After a long day of riding through open plains, they gained the cooler foothills, and the next morning the collector left Cock de Lard with the horses in a shady cedar grove and ascended a nearby ridge. There he found

> a remarkable spring that rises on the summit, from a circular hollow in the earth, eleven feet in diameter; the water springs up to from nine inches to three feet and a half from the surface, gushing up and falling in sudden jets; thence it flows in a stream down the mountain fifteen feet broad and two and a half feet deep, running with great rapidity, with a descent of a foot and half in ten, and finally disappears in a small marsh. I could find no bottom to the spring at a depth of sixty feet.

A lush thicket of currants "bearing fine fruit, much like gooseberries, as large as a musket-ball, and of delicate and superior flavor" surrounded the spring, which he named Munro's Fountain after the chief gardener at Chiswick, then doubled the compliment by dubbing the delicious currant *Ribes munroi*.

Douglas had indeed found a remarkable place, on the slopes below modern Lake Waha, the only large natural lake in the vicinity. Some time in the past a large landslide blocked the lake's outlet, and the resulting hydraulic pressure forces water through the seams of sedimentary beds intercut with basalt flows. Springs bubble up along the landslide scar, and even on bone-dry summer days these seeps support lush growths of currant and cow parsnip, elderberry and sphagnum moss.

When Douglas rendezvoused with Cock de Lard and the horses at the grove, he found that their provisions included only enough salmon for one day. With dusk approaching, the pair decided to ride back to the Clearwater that night. Cock de Lard lost his way in the open country, and only after a late-rising moon illuminated their position did they realize they had ridden an extra ten miles. When they finally straggled into camp at sunrise, Douglas found a bit of food, then collapsed in his tent.

In keeping with the pattern of interrupted sleep that typified his summer, he was awakened two hours later by a loud dispute. The interpreter, a man named Toupin, was being called a dog by one of the Nez Perce chiefs, and was responding in kind. The argument escalated until "the poor man of language had a handful of his long jet hair torn out by the roots." The traders and Nez Perce men were facing off "with their guns cocked and every bow strung" as the groggy Douglas emerged from his tent. "Aroused to take on myself the profession of a soldier," he quickly fell in line with his mates, who appeared to be outnumbered by better than two to one. John Work kept a cool head, refusing to hand over Toupin for punishment, while appealing to the Nez Perce chief to avoid bloodshed. He then passed around many inches of tobacco and patiently waited through several hours of conciliatory speeches.

Douglas spread his seeds to dry in the warm wind and watched the parlay, concluding that the tribal speakers were as persuasive as many English orators he had heard. By the end of the day, peace had been restored and the company had gained seventy-nine new horses. John Work immediately ordered two of them butchered for food.

Wickerwork

On the last day of July, Archibald McDonald bade farewell to Douglas and returned to Fort Walla Walla. He was replaced by Finan McDonald, who joined John Work and the drovers as they set out along an ancient tribal track that followed coulees carved by Ice Age floods along the Palouse River and Cow Creek. Although badger and jackrabbit holes made riding through the bunchgrass dangerous, the drive covered forty-five miles before stopping to camp. Douglas had only enough time to gather a couple of locoweeds, but the next day a predawn shower refreshed the air and slowed the pace. When the party paused for a siesta, the collector left his watch with Work and started off on a long loop that would intersect their route; after three hours he had circled all the way back to their lunch spot to find everyone still snoozing.

That evening they camped under the shade of a single cottonwood beside a stagnant pool that swarmed with frogs and tiger salamander larvae. "Water very bad," Douglas wrote, but at least a fanning breeze kept the mosquitoes away. He found a pondweed, a spatterdock, and a water lily in seed, and the next day picked up a pale monkeyflower that suited his fancy. As they approached the higher ground of the Spokane country, he welcomed the sight of ponderosa pines, and on

August 2 the party stopped for breakfast at a cool lake west of present-day Cheney, Washington. Faced with a choice of routes, they veered north to strike the Spokane River at the mouth of Latah Creek, just downstream from the waterfalls that mark the modern city of Spokane. While the drovers turned their horse herd downriver, Douglas rode upstream to admire the falls, "a perpendicular pitch of 10 feet across the whole breadth of the river."

On a 99 degree afternoon, Work and his men coaxed the horses around difficult basalt formations along the Spokane River. When darkness fell, they decided to pitch camp rather than risk traveling across the sharp rocks. Douglas, plying his own trade, searched in vain for seeds of yellow bells, but took some compensation in gathering ripe capsules of mariposa lilies.

At nine o'clock the next morning, they came abreast of Spokane House. "Old Mr. Finlay" came out to greet them as soon as they crossed the river, and this time he had better fare to offer: the late-summer chinook salmon were running. The traders ate a fine breakfast, then divided their horse herd: one group of animals would be driven to Fort Okanagan, while John Work and Finan McDonald would push the remainder up a tribal road to Fort Colvile.

After the horse brigades departed, Douglas lingered to examine an elaborate fish barrier near the confluence of the Little Spokane and Spokane rivers.

> *The barrier, which is made of willows and placed across the whole channel in an oblique direction, in order that the current which is very rapid will have less effect on it, has a small square of 35 yards enclosed on all sides with funnels of basket-work (just made in the same manner as all traps in England), and placed on the*

JACK NISBET

under side, through which the salmon passes and finds himself secure in the barrier.

He studied the way the fishermen's spears were crafted from two interlocking pieces. A sharpened bone formed the tip, which was laced tight to a pointed piece of wood; this was fastened to a long staff by a cord and fitted into a socket in such a way that it would be released when a fish was struck. Douglas watched the weapons in action.

When the spearing commences, the funnels are closed with a little brushwood. Seventeen hundred were taken this day, now two o'clock; how many may be in the snare I know not, but not once out of twelve will they miss bringing a fish to the surface on the barb.

After counting the capture of almost two thousand salmon, he continued on his way, riding among the Indian hemp plants that the Spokane people braided into the sturdy cord that secured barb to spear. The trail followed the north bank of the Spokane for some miles before veering away from the river, and Douglas apparently paused near here to admire a large fir tree that towered over the path, then pulled out his knife to carve "D. Dgls" into its trunk. That night he caught up with the horse brigade at their camp beside a stand of white-flowered currant that had attracted his attention earlier in the summer. This time he was able to gather seed from its dried fruit.

Plodding along the wetlands of the Colville Valley the next day, Douglas hesitated at the Little Pend Oreille River, where earlier in the spring two strenuous swims had brought on attacks of rheumatism. This time the water was low enough to ford on horseback. A few miles farther on, he came to the bubbling spring where, fifteen years before, David Thompson had cut cedar trees to build a plank canoe. The spring fed a deep creek that was still running fast. Douglas thought he had forded safely until "my horse on gaining the opposite shore, which is steep and slippery, threw back his head and struck me in the face and I was plunged head foremost into the river." He emerged without injury, but his seed packets and journal were thoroughly soaked. Shaking off the misfortune, he trotted into Fort Colvile with the drovers at dusk, exactly two months after he had departed in a canoe headed downstream. John Dease "cordially and hospitably entertained" the new arrivals through the evening, and Douglas was soon back to his collections, paying particular attention to the death-camas that flourished in the Spokane and Colville valleys.

As the hot weather waned, John Work offered to take Douglas east to the Flathead country, but tribal unrest there made the collector fear he would be restricted to camp. Then in mid-August Dr. McLoughlin sent word that the vessel *Dryad* was scheduled to sail for England on the first day of September. Douglas had already sent two boxes of material to Fort Vancouver for shipment to London, but now he decided he wanted to add more. When he proposed traveling downstream to Fort Vancouver, Dease protested that the river was still too high to go by boat, and arranged for a guide named Little Wolf to accompany him as far as Fort Okanagan on horseback.

Douglas packed a trunk with a bundle of dried plants and a flat twined bag "of curious workmanship," fashioned from Indian hemp and bear-grass with eagle quill decorations. He added his own small kit, "consisting of one shirt, one pair of stockings and a night-cap, and a pair of old mitts." Just as he and Little Wolf were ready to depart, a Kootenai party appeared from the north. This tribe had an ongoing dispute with Little Wolf's people, and before the afternoon was out, the opposing groups were lined up in the camp, painted for war. When one man drew his bow, John Dease stepped in and smacked him hard in the nose, which so shocked all of the combatants that they agreed to a temporary truce. Little Wolf had to devote all his attention to the ensuing negotiations, leaving Douglas stalled once again.

Dease solved the problem by recruiting a Spokane guide named Robado. Concerned with the naturalist's ragged apparel, Dease gave him some buckskin breeches and several pairs of moccasins. Upon learning that Douglas had left his battered musket at Fort Walla Walla (apparently Jaco Finlay's repairs had been only temporary), John Work loaned the collector an odd "double-barreled rifle-pistol." Apprehensive about traveling the six hundred miles between Kettle Falls and Fort Vancouver without the support of a fur trade brigade, Douglas accepted the firearm and figured that the less he had of visible value, the better off he would be.

Chasing the Dryad

When Robado and Douglas set out on August 19, neither of them could understand a word of the other's language. For the first day they followed a rocky trail south toward Chamokane Creek, camping at sundown beside a fine spring. Impatient to be on the move, Douglas arose at 2 a.m.; Robado obligingly picked his way by moonlight to a fishing encampment on the Spokane River near Little Falls. For the price of a few crumbs of tobacco, four Spokane men led the travelers downstream and ferried them across the river's mouth. Soon the two riders were out of the trees and into the sagebrush again, good horse country where they rode hard to make time. Following the south bank of the Columbia, Robado stopped at a stagnant pond that Douglas deemed unsuitable for a campsite. They plodded on to a freshwater spring, arriving at 11 p.m. after twenty-one hours on the trail. Without a twig of wood to start a fire, the collector gathered horse droppings to boil his tea; he was so exhausted that he slept in until 4 a.m.

By midmorning the next day, August 21, they had reached the "stony chasm" of Grand Coulee, which Douglas guessed must have once formed the bed of the Columbia. He found it "a truly wonderful spot, in some places eight or nine miles broad, and exhibiting such rocks in the channel as must have occasioned prodigiously grand cascades." When he measured the canyon's great breadth and height, he was visualizing the scale of the Ice Age floods; when he observed that the "plants peculiar to the rocky shores of the Columbia are to be seen here and in no intervening place," he was rec-ognizing the unique nature of Columbia Basin flora. But such lofty insights did not prevent him from attempting one of his clunky jokes: "The whole country covered with shattered stones, and I would advise those who derive pleasure from macadamized roads to come here, and I pledge myself they will find it done by Nature."

On their way through the coulee, they stopped at a spring whose surface smelled so strongly of sulfur that the horses would not touch it, even though they had been without water for twelve hours. Robado found a small pond covered with duckweed, and this the animals attacked with such fury that one tumbled in headfirst and became stuck in the mud. Douglas was cocking his pistol to put the mired horse out of its misery when he had the inspiration to jab its flank with his penknife. The steed bolted straight out of the muck and crashed onto dry land. At least the animal's thirst had been slaked; Douglas's had not: "The water was so bad that it was impossible for me to use it, and as I was more thirsty than hungry I passed the night without anything whatever."

As their trail climbed out of Grand Coulee through Barker Canyon, he found the terrain much more to his liking, and by midday they were walking into Fort Okanagan, where Archibald McDonald was on duty. Archie provided his tired friend with a clean shirt and a good supper. When Douglas explained his desire to catch the *Dryad* before she sailed, McDonald offered a small but serviceable canoe with a trusted guide. Everything would be ready, he promised, first thing the following morning. Before turning in, Douglas expressed his thanks to Robado, who had "behaved himself in every way worthy of trust," and gave him a note for John Dease reporting the success of the first leg of his journey.

Douglas and his new escort set off at seven a.m. on August 23, short of provisions but with a fresh supply of tea and sugar and a little shaving pot for a kettle. The guide's fourteen-year-old son helped paddle, and when they encountered fast water at the Methow River, the rapid-shy Douglas grabbed his papers, seeds, and blanket, and set off on foot. Halfway around the portage he watched a standing wave comb the canoe clean, leaving nothing except the paddlers and some dried meat that was wedged between thwart and gunwale. When they relaunched below the rapid, the naturalist took the stern to serve as steersman for the rest of the day, and by the time they camped near the mouth of the Entiat River, his hands were covered with blisters.

Reaching the head of Rock Island Rapids the next morning, Douglas offered a smoke from his pipe to a local man in return for piloting the canoe through the dangerous channels. After that obstacle was safely cleared and the promised pipe smoked, the Columbia swept them down to Priest Rapids, where darkness forced them to make camp midway through the long series of shoals. Upon arising early the next morning, he walked to the base of the rapids, gathering seeds along one of his favorite stretches of river while he waited for the father and son to bring the canoe down. When they failed to appear, the collector became alarmed for their safety and headed back upstream on foot, only to find his paddlers "comfortably seated in a small cove treating some of their friends to a smoke with some tobacco I had given them."

He must have goaded the pair into quick action, because by sundown they were beaching the canoe at Fort Walla Walla, where he paid them for their services with ammunition and more tobacco. Too weak to eat, Douglas "laid myself down on a heap of firewood, to be free from mosquitoes." He awoke to find that Samuel Black had placed a larger canoe and two new guides at his disposal. Precisely at noon on August 27, the new crew reached Celilo Falls. Their boat was too heavy to heft around the portage, but Douglas, to his great relief, was back among people who could converse in Chinook jargon. He cached Black's canoe and hired a smaller craft to taxi him to the head of the Dalles.

The trading bazaar there was teeming with hundreds of people, among them a man carrying the rack of a male bighorn sheep, which Douglas purchased on the spot for three musket balls and the powder to fire them. The man also wore a robe made from a *mouton gris*'s dressed skin, but for that he demanded too steep a price—the linen shirt off Douglas's back. Refusing to barter, the collector refreshed himself with a snack of hazelnuts and huckleberries, then walked the long Dalles portage trail in quest of his next ride. Along the way he met an acquaintance named Red Coat, "a valuable man, as I understood his language." Red Coat proved even more

valuable when he professed himself willing to ferry a passenger downstream the following morning. Congratulating himself on his acumen, Douglas made camp at the base of the
rapids, built a fire, and hung his coat to dry.

But few travelers of this era made it through the Columbia
Gorge without some discomfort. Before long, he was joined
by several dozen men who wished to share a smoke. Reaching
into the pocket of his jacket, he discovered that his tobacco
tin had disappeared, and concluded it had been pilfered by
one of the crowd. He clambered up on the rocks and "in their
own tongue I gave them a furious reprimand, calling them all
the low names used to each other among themselves. I told
them they saw me as only one *blanket man*, but I was more
than that. I was the *grass man*, and was not afraid." The Grass
Man never did see his tobacco tin again, but he did sleep
undisturbed.

Stiff headwinds hampered Douglas and Red Coat's downstream run to the Cascades of the Columbia, and when they
finally reached the top of the portage on August 29, the collector immediately headed for the lodge of Chumtalia, his
gracious host of the previous summer. In concord with everyone else on the river at that time of year, Chumtalia was
busy processing fish, but he provided Douglas with a sumptuous meal of fresh salmon and huckleberries, laid out on a
tule mat. He also proffered a large canoe, with his brother
and a nephew as paddlers, to transport the Grass Man on the
last lap of his race in style. When Chumtalia's vessel docked
at Fort Vancouver at lunchtime the following day, Douglas
acknowledged "the gratification of . . . traversing nearly eight
hundred miles of the Columbia Valley in twelve days and
unattended by a single person, my Indian guides excepted."
His estimate of river miles was a couple hundred on the long

side, but his journey did stand as an ambitious exploit for any single traveler.

As the bedraggled collector strolled across the low plain between the river and Fort Vancouver on August 30, observers thought that the shoeless man approaching the post, dressed in deerskin trousers and a battered straw hat, must be the lone survivor of a boat mishap on the river. After recognizing their resident naturalist, whose "careworn visage had some appearance of escaping from the gates of death," they were greatly relieved.

For his part, Douglas was relieved to find Captain Davidson of the *Dryad* still at the post, preparing to sail for England the very next day. Dr. McLoughlin produced boxes for the collector's new material, and after a hurried round of packing and letter-writing, his cartons and mail were placed in the captain's hands. By then the Grass Man "stood in need of a little ease," and after bidding adieu to his friends who were returning to London on the *Dryad*, he made a more relaxed walk from the river to the post, excited to see what the fall might bring.

VII.
THE PERFECT ENTHUSIAST
FALL 1826–SPRING 1827

Looking for the Umpqua

Douglas spent the first two weeks of September at Fort Vancouver, assembling selections of wood, gum, and bark from the adjoining forest. As if emulating the local tree squirrels, he stored away many mature cones of the conifer called *Pinus taxifolia*, a tree with needles like a yew, bark like a hemlock, the pendulant cones of a spruce, and the dense reddish wood of a larch. Two twenty-foot logs of that species were on the way to England in the hold of the *Dryad*. Months later, John Lindley would take receipt of them for the Horticultural Society, and soon thereafter British timber men would realize the potential of the tree today known worldwide as Douglas fir.

Among the traders assembled at the post that September was Alexander Roderick McLeod, whose company Douglas had enjoyed the previous winter. McLeod was preparing an expedition to the rugged Umpqua drainage, the source of the unforgettable outsized pine cones. When John McLoughlin

117

suggested that Douglas join the party, he "could not allow this favourable opportunity to escape" and began assembling his kit. Because his musket had finally given up the ghost, he purchased a new one at the company store for two pounds, and filled a small copper kettle with trinkets and tobacco for trade. He limited his extra clothing to two sturdy shirts, but did indulge in an extra blanket and a tent in anticipation of rainy weather. After making room for six quires (144 sheets) of paper, "requisite for what I call My Business," on September 20 he joined twelve men who were ferrying supplies up the Willamette in a large bateau. Two days later, they reached McLeod's base camp near a Kalapuya Indian village known as Champoeg (near the modern Oregon state park of the same name).

The encampment housed over thirty people, including two Indian guides, eleven French Canadians, five Hawaiians, and several tribal wives and mixed-blood children. The party would be switching to horseback at this point, and Dr. McLoughlin had generously sent up one of his "finest and most powerful horses" for Douglas's use. After a few days spent arranging their loads, the pack train headed up the Willamette Valley, across undulating plains spotted with solitary oaks and scattered conifers.

Over the next week the party traversed a fire-scorched landscape that proved "highly unfavourable to botanizing." When Douglas inquired about the extensive burning, some of the Kalapuyas traveling with the party explained that their people set fire to the plains each fall so as to lure deer into unburned areas where they could be easily hunted. Others added that it was "done in order that they might the better find wild honey and grasshoppers, which both serve as articles of winter food." Like the Spokanes and other Plateau tribes, the Kalapuya people, taking into account subtle cues of season, weather cycle, and habitat, set controlled fires for a variety of outcomes. Some were intended to create open space and stimulate the growth of specific shrubs. Smaller burns were used to toast tarweed seeds or concentrate insects, such as the grasshoppers that the Kalapuyas mentioned to Douglas. But their reference to wild honey is puzzling, since honeybees are not native to the Northwest and had not yet been imported from Europe. Douglas might have confused the word for *bee* with that for another insect such as a paper wasp or yellowjacket, whose larvae-filled nests, when roasted by fire, provided a source of protein for some valley tribes.

As far as McLeod's horse brigade was concerned, the charred country created only relentless hardship. They often trudged many extra miles in search of fodder for the horses. "Pasture is rarely found in the course of this day," the agent wrote; ". . . everywhere the fire destroyed all the grass." Sharp burned stumps of ninebark and reed grass tore the men's shoes and cut their feet. Rations were short. "Thus we live literally hand to mouth," Douglas wrote, "the hunters all declaring that they never knew the animals of all kinds to be so scarce . . . which is attributable to the great extent of country that has been burned."

On Sunday, October 1, after traveling eighteen miles, the party made camp on the banks of a small stream. The product of a strict Presbyterian upbringing, Douglas was adjusting to the traders' more casual approach to the Sabbath, recognized at Fort Vancouver by a short service and on the trail "only by the people changing their linen, while a few of the men who could read perused religious tracts." Having presumably taken care of both those observances, the collector struck out at dusk to search for game with hunter John Kennedy. In a blackened flat a short distance from the camp, they happened upon a large yellow-jacket nest, apparently torn from a tree and dragged into the open by a bear. While the collector pondered the hive, Kennedy spotted a large grizzly entering a hummock of brush not two hundred yards away. The two men decided to leave the beast unmolested in the gathering darkness, but early the next morning Douglas went in pursuit, thinking of how welcome such a quantity of meat would be.

He found no sign of the bear, and spent most of the day rambling through a deep ravine filled with bracken, golden-rod, and thistle. Writing in his journal that evening, he complained of his aching feet and cut toes, but failed to mention the birth of a baby girl to one of the women in the party, whose "indisposition" had caused a delay on the trail.

During the first week of a rainy October, Douglas spotted another iconic creature of the Northwest, the California condor: "The Large Buzzard, so common on the Columbia, is also plentiful here; saw nine in one flock." Near present-day Albany, McLeod pointed out the distinctive cone of Mount Jefferson, framed by two of the snow-covered Sisters; the tribes called the country beyond these peaks the "Clamite" (Klamath). Finan McDonald, who had explored the area the previous year, had told Douglas that the large-coned pines could be found there, and the collector considered a side visit to that area. But at the moment he was tied to McLeod's party, and his sights were set on the Umpqua.

As Douglas gazed toward the Klamath and dreamed of pines, Jean Baptiste McKay, the man who had brought him the gargantuan cone the previous winter, appeared on the scene with two Iroquois trappers. McKay had been hunting in the area and carried the skin of a large female grizzly bear he had killed a few days earlier. When Douglas spied the thick pelt, he realized it would be ideal "to use as an under robe to lie on, as the cold dew from the grass is very prejudicial to my health." He proffered a blanket and some tobacco in exchange, and placed an order for a matching male pelt.

By chance, at that moment just such a boar grizzly was charging John Kennedy in a small oak grove near camp. When a musket shot had no effect on the bear, Kennedy scrambled up a small oak. The beast clawed at his coat and breeches, but "fortunately his clothing was so old that it gave way, or he must have perished."

The horse brigade left the open country of the Willamette Valley and entered the hilly, wooded terrain south of present-day Eugene on October 9. Douglas, who had often lamented during the march across the scorched valley that "nothing new came under my notice," was rewarded with the sight of a golden chinquapin.

> *Its rich varied foliage, quivering in the wind, clothed to the very roots with wide-spreading branches, and standing alone on the dry knolls or on the crevices of rocks, gives a tint to the general appearance of American vegetation of more than ordinary beauty.*

The collector had no idea whether this princely tree was a new species, but he felt certain it was rare. Hoping to find an example of its fruit, he explored the hillside, and after a laborious search found a single tree bearing the prickly golden husks that shelter its seeds.

Descending the hummock at dusk, he stumbled through an abundance of the sticky lily called false hellebore, then made his way down a serpentine stream to rendezvous with the brigade in a small cove, where the hunters had brought in a bull elk. Like the coastal deer, it was small for its kind, weighing only about five hundred pounds. Although Douglas found its meat lean and tough, he was impressed by its five-pronged antlers and measured them carefully, wondering if this might be the same species he had seen on the Duke of Devonshire's estate, adjacent to the Chiswick gardens.

The next day Douglas shadowed the party on foot, plodding uphill through heavy rain. The grand firs of the lower slopes gave way to enormous hemlocks. Stunted madrones with large trunks had been splintered by bears clambering after the succulent orange-red fruit. Winding a circuitous route around the slow-moving horses, he pressed woodland shrubs such as salal, ninebark, manzanita, Oregon grape, hazelnut, and huckleberry. All the while, he thrashed through great thickets of bracken fern, eight to ten feet tall, laced with blackberries and vetch, which "rendered it so fatiguing that every three hundred or four hundred yards called for a rest."

That evening, Alexander McLeod decided the horses were also in need of a respite, and the party remained in camp for two more rainy days. After drying his soaked clothing and blanket, Douglas stalked his first mountain quail, "a most curious little bird," with a purple-red throat patch and black feathered topknot. He brought down one out of a flock of five, but his shot ripped off the beak and one leg, and he spent a soggy afternoon in unsuccessful pursuit of a more perfect one to send home.

On Friday, October 13, the party crossed the divide between the Willamette and Umpqua drainages at the headwaters of Pass Creek (near Divide, Oregon). When McKay and McLeod shot a pair of white-tailed deer, Douglas carefully measured every dimension of the female and tied one of the buck's antlers to his knapsack. McKay roasted venison steaks with an infusion of sugar and wild mint, then served the feast on a plate of salal leaves. The collector ate his meal with a spoon fashioned from the horn of a *mouton gris*.

The brigade continued to move south, hacking their way through thickets of vine maple on wearyingly steep slopes. During the afternoon, they reached the mouth of Pass Creek (near present-day Drain, Oregon) and turned down Elk Creek. Rain kept everyone in camp for the next two days, during which Douglas arranged his gun and read some old newspapers. Back on the trail on October 16, he scouted ahead on foot with McLeod and McKay, hacking out a path for the pack train on a roller-coaster route strewn with fallen trees. The wet ground "rendered the footing for the poor horses very bad; several fell and rolled on the hills and were arrested by trees, stumps, and brushwood." Fearful that his horse might suffer such an accident, Douglas bundled his plants inside a small bear skin and toted them on his back. It must have been an awkward load, but that didn't stop him from adding one more curiosity: "Found one shed horn of a Black-tailed Deer: the temptation was too great, so I tied it on the bear-skin bundle."

As the bushwhackers cut their way through the deep valley along Elk Creek, the air was filled with a peppery aroma emanating from a stand of California laurel trees, known by today's Umpqua residents as Oregon myrtle. Douglas entertained himself by rubbing the leaves between his hands until he sneezed. Some of the hunters mentioned that they made tea from the tree's bark, but he was more interested in seeds. He tried shinnying up to a fruit-laden branch; when the smooth white bark proved too slippery to climb, he resolved to chop down the tree, "which was done at the expense of my hands well blistered." He tucked the seeds into his bear skin, predicting that "this fine tree will ere long become an inmate of English gardens, and may even be useful in medicine, and afford a perfume."

Stopping often for a smoke to stem their fatigue, the advance team finally reached the Umpqua River late that afternoon. While McKay and McLeod hunted, Douglas kindled a fire and refreshed himself with a dip in the river. Observing Sitka spruce and false azalea (rusty menziesia) along the shore, he deduced that the ocean must be within thirty or forty miles, since those two species "always keep along the skirts of the sea."

During the evening, a few of the men and horses straggled in with word that the main party had made camp some distance back. Douglas's mount, laden with his blanket and grizzly rug, was among the laggards. Unable to sleep on the cold ground, he huddled beside the campfire until 2 a.m., when McLeod awoke and insisted on sitting by the fire while Douglas took a turn on his buffalo robe.

When the tired horses of the main brigade finally made it in the next day, Douglas saw his fears confirmed.

> *The baggage which mine carried was almost destroyed by the poor beasts' repeated falls; the tin box containing my notebook bruised quite out of shape, its sides bent together—a small case of preserving-powder quite spoiled,—and my only shirt reduced, by the chafing to the state of surgeon's lint. I congratulated myself exceedingly on not having trusted my papers of plants to the same conveyance, but carried them on my back.*

McLeod decided to remain encamped at the mouth of Elk Creek for a few days to rest his pack animals, and Douglas seized the opportunity to search for the fabled pines.

Centrenose and Son

The reputed home of the big-coned trees lay in a mountainous area to the south. Accompanied by an eighteen-year-old Indian from McLeod's crew who spoke both Chinook jargon and Umpqua, Douglas had not proceeded far in that direction when he reached two Indian lodges, where "the children on seeing me ran with indescribable fear." A man and a woman emerged, and after discerning Douglas's "friendly disposition," beckoned several other inhabitants, mostly women, into the open. Douglas's young companion proved to be less than fluent in the Upper Umpqua dialect, but the women of the camp spoke the universal language of hospitality, and soon the two travelers were feasting on hazelnuts, camas roots, a gruel made from ground myrtle nuts, and a tea brewed from the leaves and shoots of the same tree. Douglas was able to communicate his desire for some nuts of this "smelling-tree," and happily paid for them with beads, brass rings, and tobacco.

As they dined, Douglas admired the untanned deer-hide breeches of the men and the woven cedar-bark petticoats of the women. His companion spoke enough Umpqua to learn that the lodges belonged to Centrenose, a principal chief of the Upper Umpquas, who was off hunting. After the meal, the villagers ferried their guests across the river in a large canoe, coaxing their horses to swim alongside, then pointed out a trail that cut across a long oxbow, shortening the next stage of their journey.

At the end of the shortcut, the pair were once again faced with crossing the swift-running Umpqua. They decided to build a raft, but after considerable toil, found it too small to support them. With nothing to show for his efforts except painfully blistered palms, Douglas sent his guide back to McLeod's camp to seek help. Although his hands were too sore to wield a hatchet, the naturalist was able to squeeze a trigger and soon wounded a large deer. He gave chase,

> but in the eagerness of pursuit, fell into a deep gulley among a quantity of dead wood, and lay there stunned, as I found by my watch when I recovered, nearly five hours.

Five Indian men who were hunting in the area stumbled upon the scene and quickly extricated the mud-covered collector. Despite his injury, Douglas had the wherewithal to lead them back to his camp and offer them some venison left over from the night before, "which gave me more pleasure than I can describe." His rescuers loaded his equipment on his horse, while the naturalist "crept along by the help of a stick and my gun." Near Centrenose's village the slow cavalcade encountered John Kennedy, on his way to help Douglas build a larger raft. Kennedy hoisted the limping collector on his own horse and escorted him back to the Elk Creek camp. After bleeding his own left foot—at this time, phlebotomy was thought to cure all ills—Douglas felt "somewhat relieved."

The following morning he bathed in the river and felt better still. McLeod was on the move again, heading downriver, and Douglas limped along at the rear of the party for a rough ten miles. Afterward, he declared himself "much improved, but still stiff as if I had been undergoing great labour." Fortunately, the next day's march was much shorter, for within five miles they reached tidewater and pitched camp.

McLeod sent a messenger to an Indian village in the river's delta, and soon twenty large dugouts of the Lower Umpqua tribe appeared. These people spoke a different language than their neighbors upriver and followed lifeways similar to the Lower Chinooks around Baker Bay. The coastal band presented the traders with a yard-long steelhead they had speared in the estuary; Douglas was surprised to learn that they fished only with spears, never with the nets so common among the Columbia River tribes.

When the Lower Umpqua party returned the next morning with more fresh steelhead there were women along, their lower faces and jaws tattooed with dark lines and spots. Douglas was still digesting his excellent breakfast and pondering different standards of beauty when McLeod shot a large black-tailed deer as it grazed among the horses near camp. Douglas immediately set to work with a measuring tape and pencil, noting the black ring around its nose and details of its antlers, tail, mane, back, and belly. Comparing his impression of a white-tailed deer dissected earlier on the trip, he concluded that the black-tailed variety was "a much larger animal than the Long-tailed, at least a fifth larger, and assuredly a very distinct species."

When he questioned the hunters about places the deer were usually found, one of them produced a snare sometimes used to capture the animals. Though it was no thicker than a little finger, the hunters assured him that the cord was "strong enough to secure the largest Buffalo and the Elk." Upon examining the snare, Douglas saw that it was woven from the leaves of a small iris that thrived in the moist rich grounds along the lower river. He was regretting that it was too late in the season to collect any seeds from these iris when Centrenose and some of his family arrived in camp. McLeod kindly employed one of the chief's sons to accompany Douglas to the interior to continue his quest for the elusive pine, and by midmorning on October 23 the collector was headed back upriver, his baggage loaded on one horse and Centrenose's son atop a second.

After a brief stop at his guide's home village, they reached the site of Douglas's ill-fated raft construction. Centrenose's son lit a signal fire, and soon a canoeist appeared and ferried the travelers across the river. That evening the collector found himself in a comfortable village "where the kind inhabitants kindled my fire, some brought me nuts, another salmon trout, a third water from the river to drink." The next day, however, was not so easy. Deep gullies, fallen timber, and impenetrable underbrush obstructed their progress. Rain fell in torrents. Camped near the present town of Cleveland, Douglas consumed the last of his venison with a few ounces of rice and realized he would be wise to limit himself to one meal a day.

When night fell, conditions continued to deteriorate. A fierce wind blew out the fire, and "to add to my miseries, the tent was blown down about my ears, so that I lay till daylight, rolled in my wet blanket, on *Pteris aquiline* [fern fronds], with the drenched tent piled above me." Gigantic trees crashed all around. Echoing peals of thunder and zigzagging flashes of lightning added terror to discomfort. The horses cowered near his bedroll, "unable to endure the violence of the storm without craving of my protection, which they did by hanging their heads over me and neighing."

The storm finally abated, and the sun rose on a clear but frigid day. Douglas kindled a fire to dry his clothes, taking a "fume of tobacco" while scrubbing his frozen limbs with his handkerchief until he could no longer endure the pain. He was still shivering when he got under way in midmorning, then "was seized with a severe headache and pain in the stomach, with giddiness and dimness of sight." His medicine kit contained a few grains of calomel—a powdery, sweet paste of mercurous chloride commonly used as a purgative and fungicide in those days. Perhaps aware that the damaging effects of the mercury might outweigh its benefits, he decided to take the calomel only as a last resort. Instead, he "bolted forward with renewed vigor" and covered thirteen miles before stopping at a small encampment of Indians, who offered a meal of salmon that Douglas judged barely edible but was grateful to obtain. His guide pitched camp earlier than usual, and after hanging all their clothing and bedding to dry, Douglas confided the travails of the previous twenty-four hours to his journal, then added a philosophical postscript:

> *When my people in England are made acquainted with my travels, they may perhaps think I have told them nothing but my miseries. That may be very correct, but I now know that such objects as I am in quest of are not obtained without a share of labour, anxiety of mind, and sometimes risk of personal safety.*

The object of pines remained uppermost in his mind. The next morning, leaving Centrenose's son to tend the horses, he headed south toward a promising ridge (near present-day Roseburg, Oregon). He had been walking about an hour when he met a single Indian, who strung his bow and slipped a raccoon-skin wrist guard onto his left hand as soon as he sighted the stranger. Douglas calmly laid his musket on the ground and beckoned the man closer, "being quite satisfied that this conduct was prompted by fear, and not by hostile intentions, the poor fellow having probably never seen such a being as myself before." Soon they were smoking together, and when the naturalist pulled out a pencil and sketched a distinctive cone, the man pointed south and cheerfully set off in that direction. Following his new acquaintance, Douglas finally reached a grove of immense trees. He set to work immediately.

New or strange things seldom fail to make great impressions, and at first we are liable to over-rate them; and lest I should never see my friends to tell them verbally of this most beautiful and immensely large tree, I now state the dimensions of the largest one I could find among several that had been blown down by the wind.

The base of the fallen tree he measured had a circumference of fifty-seven feet, nine inches. When alive, the giant had stood two hundred and fifteen feet tall. Its trunk had been clear of branches for two-thirds of its great height, with a remarkably straight stature. Globules of bright amber gum oozed from its uncommonly smooth bark.

But the downed tree bore no fruit. The only cones to be seen were high in the tallest of the living trees, hanging from the points of pendulous branches "like small sugarloaves in a grocer's shop." There was no way Douglas could climb to those lofty branches, or hew through the massive trunk. Taking the logical alternative, he loaded his musket and "endeavored to knock off the cones by firing at them with ball." When he lowered his gaze back to ground level a few minutes later, he saw that the sound of his gun "had brought eight Indians, all of them painted with red earth." Armed with bone spears, flint knives, and bows and arrows, the group did not appear particularly friendly, so Douglas set about ingratiating himself. He offered a pipe of tobacco, and they all sat down to smoke. But when one of the men began stringing his bow while another whetted his knife with a pair of wooden pincers, Douglas quickly stood aside, cocked his musket, and pulled a pistol from his belt. "As much as possible I endeavored to preserve my coolness, and thus we stood looking at one another without making any movement or uttering a word." After a few minutes, the leader signaled for another smoke. Douglas indicated that he would exchange more tobacco for a quantity of cones, and the warriors scattered to search the forest.

No sooner were the eight men out of sight than a skittish Douglas gathered the three cones he had shot down, grabbed samples of branches and needles, and along with his friendly companion, "made the quickest possible retreat." Not certain how far he could trust even that fellow, he parted ways with the man with the raccoon-skin wrist guard a good distance from his camp "lest he should betray me."

Before nightfall he had time to finger the twigs he had saved, to count the five bright green needles in each sheath, and to ponder the shortness of both sheath and needle. He measured his three precious cones—fourteen inches, fourteen inches, thirteen inches—and made sure that each was filled with fine, ripe seed.

That night, unable to communicate his worries about the hostile Indians to Centrenose's son, Douglas lay beneath his blanket with his gun cocked at his side. A "Columbian candle, namely an ignited piece of rosiny wood," illuminated his journal, and his thoughts alternated between fear of an attack and the tidbits he had learned about "the tree which nearly cost me so dear." Tribal people set fires around the trees to obtain the heated resin, "a substance which, I am almost afraid to say, is *sugar*," Douglas wrote, adding that he hoped to bring some of the sap back to England to be properly analyzed. As for the cones of the tree now called the sugar pine, they were "gathered by the Indians, roasted on the embers, quartered, and the seeds shaken out, which are then dried before the fire and pounded into a sort of flour, and sometimes eaten round"—that is to say, whole.

Despite his anxiety, he eventually dozed off and was sleeping soundly when a loud shriek started him awake. "Thinking the Indians I saw yesterday had found me out," he sprang to his feet and found his guide in a state of terror— he had gone out before dawn to spear fish by torchlight and had been charged by a grizzly. After waiting for a little more daylight, Douglas mounted his trusty steed and tracked an adult bear and two cubs to the base of a large oak where they were feeding on acorns. He shot one cub, then fired at the sow, who fled into the brush with her remaining cub. Douglas bequeathed the dead cub's carcass to Centrenose's son, "who seemed to lay great store by it."

At the young man's village the next evening, there was no extra food on hand because muddy runoff from days of rain had disrupted fishing. Douglas ate the last of his rice, then paid his guide and set out alone early the next morning. On a slippery hillside, his horse stumbled and rolled down a steep slope. Bouncing over dead logs and large stones, the animal "would have been inevitably dashed to pieces in the river, had he not been arrested by being wedged fast between two large trees that were lying across the hill." Douglas hobbled its legs and cinched its head so it couldn't struggle, then chopped the trees away with his hatchet. To his great relief, the animal was only slightly injured. "I felt over this occasion much, for I got him from Mr. McLoughlin and it was his favourite horse."

A Passable Hunter

Douglas grasped the jaded horse's bridle and led it downriver through pounding rain, determined to reach McLeod's base camp near the mouth of the Umpqua before dark. Michel Laframboise, the party's interpreter, was the only person on hand, but he quickly pitched a tent for the weary traveler, served him some spirits, and made a refreshing basin of tea, all the while relating troublesome encounters with the local tribe since McLeod's departure for an excursion down the coast a few days earlier.

Both men remained on watch through the night, and when Laframboise detected suspicious movements in the shadows, they stoked the fire and hid nearby. At first gray light they perceived a party of fifteen men crawling toward the tents. Laframboise and Douglas fired their muskets in the air, then watched in relief as the intruders dashed away. After a breakfast of dried salmon and tea, the thoroughly soaked collector retired to dry his clothes and "sat in my tent with a small fire before the door the whole day."

The weather remained wet and the situation tense. Douglas gathered pine branches and ferns for bedding, then again helped Laframboise keep watch all night. Although no trouble surfaced, fatigue and anxiety undermined the collector's stamina. His spirits improved when Jean Baptiste McKay returned on November 1. "It is a relief to find our little party becoming stronger, and the addition of McKay is peculiarly welcome, as he is so good a hunter, that he will procure us fresh food." By noon the next day McKay had brought in a white-tailed deer, and Douglas, "glad to stand cook," prepared a large kettle of venison and rice soup. They were just sitting down to eat when thirteen voyageurs who had been trapping in the area paddled into view and joined their company. Although he often found fur trade society "uncouth," Douglas very much enjoyed the evening that followed. "We have been entertaining one another, in turns, with accounts of our chase and other adventures, and I find that I stand high among them as a marksman and passable as a hunter."

When McLeod returned to the base camp on November 4, he filled Douglas's ears with reports of groves of sweet-smelling laurel trees and friendly Indians. The agent planned to continue his explorations over the winter, but correctly guessed that Douglas would prefer to return to Fort Vancouver with John Kennedy and a French Canadian named Fannaux, who would be delivering pelts collected by the trappers.

On November 7, with his baggage wrapped in bear skins for protection from the rain, Douglas expressed his gratitude to his congenial host of the past two months, then jotted a note in the margin of his journal: "Recollect on your arrival in London to get him a good rifle gun as a present." McLeod, for his part, regretted that he could provide only a few dried steelhead and a small quantity of Indian corn and rice for the journey.

The next afternoon Douglas, Kennedy, and Fannaux reached their previous encampment at the juncture of Elk Creek, where the collector retrieved the antlers he had cached weeks earlier. When he awoke the next morning, he had gained another bit of baggage: one of McKay's dogs, who apparently had developed an affection for Douglas, had tracked him to camp and taken up "his accustomed place, asleep at my feet." He let the mutt tag along, explaining by way of excuse that he had no way of sending him back.

Days of heavy rain had rendered the trail up Elk Creek so slippery that the horses could barely negotiate some of the hills. Matching their desultory pace, the men crept along under a rare clear sky, but had no luck at all hunting. When they camped on the evening of November 9, Kennedy and Fannaux whipped up a stew of pounded camas roots, which both declared to be very fine. Douglas's stomach did not agree—two spoonfuls of the rich, sweet concoction kept him awake all night, and "long ere day I was up by the fire and anxiously wishing for the morning, and certainly wished for a little tea, the greatest and best of comforts."

There would be no tea and little comfort for some time. The long hill that marked the Umpqua-Willamette divide was a slither of mud. All day the trio shot only a single goose, scanty fare for three mouths. Darkness caught them short of any decent pasturage for the horses, and when they finally made camp no one could get a fire going. As the three men huddled in the tent, Douglas voiced his opinion that they had wandered off course. Kennedy and Fannaux disagreed. Cold, wet, and hungry, each man clung tenaciously to his opinion. Douglas admitted that "on such occasions I am very liable to become fretful."

Early the next morning, after arranging to meet Kennedy and Fannaux with the horses at a lake several miles farther on, he set off on foot to hunt. Within an hour, he realized that he was lost. He backtracked to the campsite, and was starting off on a new course when he met Kennedy, who had become concerned when he reached the lake and did not find Douglas. As they skirted a wide wetland, Douglas proposed that they split up to increase their chances at hunting. Kennedy agreed, but not without giving the naturalist "a strict caution about going astray a second time."

Of course that is exactly what happened. Douglas quickly shot three geese and a duck and laid them on the ground beside his gun sheath and hat while he went in pursuit of more game. After winging and retrieving another bird, he could not locate his hat and other bounty. With darkness falling, he gave up and headed for the lake; Kennedy and Fannaux signaled him in by firing their guns. As he approached the camp, Douglas heard a flutter of wings and took a blind shot straight overhead. Strangely enough, he was rewarded with a duck falling from the sky. His two companions further rewarded him with intense ribbing.

> I was hailed to the camp with "Be seated at the fire, Sir," and then laughed at for losing myself in the morning, my game and other property in the evening. There is a curious feeling among voyageurs. One who complains of hunger or indeed of hardship of any description, things that in any other country would be termed extreme misery, is hooted and brow-beaten by the whole party as a pork-eater or a young voyageur, as they term it.

A pork-eater was one thing Douglas did not want to be called, and it was with renewed good humor that he sat down to dine on his duck. He rose extra early the next morning to search for his lost belongings, and had the good fortune to find not only his hat and birds, but also an additional goose he had killed without realizing it. "Today could afford myself and people breakfast!" he boasted.

Back in the Willamette Valley at last, the three men swished through lush grasses that had sprung up from the late summer burns. On November 13, they camped beside a small lake populated with thick shoals of ducks. Their gun-fire attracted several friendly Kalapuyas, who stayed for supper and led them to the nearest river crossing. Two days later Douglas's party reached the Santiam River. The current was too swift for a raft, so Fannaux took to the water astride his horse, while Douglas chose to swim. Halfway across, Fannaux and his horse parted ways, and the two men reached the shore in equal states of wetness. Not so amusing was the condition of Douglas's specimens, and he spent the rest of the day sitting before a fire, patiently drying everything as best he could.

The riders were descending the west bank of the Willamette when they met a Kalapuya chief named Tochty, or Pretty, who pointed out a more direct route to Champoeg. By this time their horses were so weak that any shortcut was welcome. The nights were growing colder, and on the morning of November 17, Douglas crept out in quest of game, but "the crisping noise among the frozen grass" alerted the deer to his presence before he could get close enough to shoot. Spotting two Kalapuya lodges near the river, he thought he might be able to purchase some breakfast. The inhabitants received him kindly, and although they were short of food themselves, one of the women presented him with a small piece of deer rump.

> *The greater part of it was only the bare vertebra, which she pounded with two stones and placed it in a basket-work kettle among water and steamed it by throwing red hot stones in it and covering it over with a close mat until done. On this, with a few hard nuts and roots of Quamash, I made a good breakfast. After paying my expenses with a few balls and shots of powder, and a few beads, I resumed my walk.*

A five-mile stroll brought him to the appointed rendezvous at Champoeg, where the men pastured their horses and finished their journey by canoe. Paddling downriver, they met a company brigade en route to the Umpqua who provided them with a lavish supper of venison, potatoes, and a basin of fresh-brewed tea. Douglas had been sorely missing his favorite beverage, but "having had no tea for some time before, it prevented me from sleeping." This meant that at 4 a.m. he was wide awake, ready to embark in a large canoe that delivered him to Fort Vancouver a little after dark on the evening of November 19.

Searching for *Sewelel*

While Douglas was exploring the Umpqua, the cross-continent express from Hudson Bay had reached Fort Vancouver, bringing letters from his brother John and various cohorts. The collector read and reread each one as he soaked his swollen ankles and repaired his shoes, then acquainted himself with several new arrivals at the post.

Among them was George Barnston, who had been born in Edinburgh within a few months of Douglas and had trained as a surveyor and army engineer, then disappointed his family by signing on with the North West Company around 1820. Governor Simpson declared his personality to be "touchy . . . and so much afflicted with melancholy or despondency, that it is feared that his nerves or mind is afflicted."

Apparently Douglas's company provided a restorative tonic for Barnston, who found "the man of science to be one of the heartiest, happiest mortals in our little society." Watching the naturalist eagerly pore over his packet of mail, Barnston wrote that the letters from home "brought out in full glow the warm effusions of a pure and happy heart." Over the winter of 1826–27 at Fort Vancouver, Douglas infected Barnston with a passion for natural history, while Barnston nurtured Douglas's interest in practical surveying. "In all that pertained to nature and science," Barnston wrote, "he was a perfect enthusiast." The two men would correspond for the rest of Douglas's life, sharing news of old mates, sightings of flora and fauna, and painstakingly observed survey coordinates.

During the first week of December, Douglas "resolved on visiting the ocean in quest of *Fuci* [seaweeds], shells, and anything that might present itself to my view." Accompanied by two Indian paddlers and one of McLoughlin's men, he was camped at the abandoned Fort George two days later. As they crossed to Baker Bay the next morning, the wind began to rise, as if on cue. Douglas pitched his tent on the beach above the high tide line and crawled inside, but this was no ordinary storm; before dawn, a surging tide flung piles of driftwood ashore, crushing the canoe and forcing him to strike his tent and retreat into the woods. Undaunted, he waited for the gale to abate, then followed the trail that led north to Willapa Bay.

There his acquaintance Cockqua greeted him "with that hospitality for which he is justly noted," regretting that he had only dried salmon and salal berries to offer for food. The fish tasted like "rotten dry pine bark" to Douglas and brought on a violent attack of diarrhea that thoroughly disabled him for four days. Fearful that he might have contracted dysentery, he struggled back to the Columbia. Paddling upriver, he stopped to buy sturgeon at a tribal encampment near Oak Point, but was totally miffed by the price: "a handkerchief off my neck for one small bit, and seven buttons off my coat for a second bit of the same size." It was a discouraged collector who trudged up the hill to Fort Vancouver on Christmas Day, "having gleaned less than any journey I have had in the country." By the turn of the new year, company food had helped him recover, and he had convinced the blacksmith to fashion him a rock hammer—probably a tool with a pointed head that he could also use for digging plant roots.

The first two months of 1827 proved tedious in the extreme, full of soft snow and little activity beyond socializing at the fort. Late in February, a welcome diversion occurred when one of the hunters brought a fresh condor carcass to the post. George Barnston and his mates "all had a hearty laugh at the eagerness with which the Botanist pounced upon it." When Douglas attempted to measure the bird's wingspan, collector and condor became completely enfolded in an awkward embrace. After finally determining that the wings stretched a full nine feet from tip to tip, Douglas ordered the hunters to take the creature immediately to his quarters so he could skin it.

Douglas was continually adding to his store of knowledge on the great vultures. On several occasions he had observed their voracious eating habits and magnificent soaring abilities, as well as their hangdog roosting posture on rainy mornings. He studied their role as carrion eaters, noting that "in no instance will they attack any living animal unless it be wounded and unable to walk." Over the winter, he and Barnston, who shared his fascination, marveled at the speed with which condors would respond to the death of any of the company's livestock—it seemed as though the moment an animal foundered, the birds began to circle. After watching the vultures drag carrion from beneath thick brush in the forest where they could not possibly have spotted it, Barnston deduced "that their sight as well as sense of smelling is very acute"—a conclusion with which modern scientists, after much experimentation of their own, have come to agree.

Douglas pursued every source of information about condor behavior, and much of the data he gathered, such as their presence in the Snake River country, was accurate. But due either to the overactive imaginations of some of the hunters he interviewed, or to his own limited French, the naturalist confused several aspects of their nesting behavior.

They build their nests in the most secret and impenetrable parts of the pine forest, invariably selecting the loftiest trees that overhang precipices on the deepest and least accessible parts of the mountain vallies. The nest is large, composed of strong thorny twigs and grass, in every way similar to that of the eagle tribe, but more slovenly constructed. Eggs two, nearly spherical, about the size of those of a goose, jet black.

Although several Pacific Northwest tribes tell stories about young condors, no positive evidence of their breeding within the region has yet been uncovered. Nor is there any evidence that they nest in trees. In California, the birds lay a single egg of a mottled brown color on a cliff ledge with no nesting material whatever.

In early March 1827, Douglas decided to make one more collecting visit to the coast. His "object was to procure the little animal which forms their robe." This particular little animal was the aplodontia, or mountain beaver, an ancient, tailless rodent endemic to the Northwest coast. A bit larger than muskrats, mountain beaver live underground in a maze of burrows, eating succulent roots. They are nocturnal and notoriously elusive. During the winter that Meriwether Lewis spent at Fort Clatsop, he tried to learn about the curious pelts from which the coastal people fashioned cloaks. "I have indeavoured in many instances to make the indians sensible how anxious I was to obtain one of these animals entire without being skined," the captain wrote. "Offered them considerable rewards to furnish me with one, but have not been able to make them comprehend me." Lewis never did see a live mountain beaver, but he did record a reasonable approximation of the Chinook word for the animal: *sewelel*. He also purchased a very fine catskin coat lined with *sewelel* fur.

Because he had read Lewis's account, Douglas would have been on the lookout for *sewelel* from the moment he landed at Baker Bay. He would have seen their burrows when he crashed through the salal on Cape Disappointment or searched for edible roots near Cockqua's village. He had purchased a fine chestnut-colored robe (sewn from twenty-seven hides), but longed to take an entire specimen back to London. Thinking he had a good chance of obtaining a *sewelel* near Cockqua's village, he set off in company with Edward Ermatinger, a clerk who had his own reasons for joining Douglas's search—if these *sewelel* pelts proved to be valuable, it would be worth his while to develop a dependable Chinook supplier.

Neither man achieved their goal on this March journey, because Cockqua was engaged in funeral ceremonies for one of his cousins. He regretted that the death prevented him from offering the visitors any assistance, but did promise to gather some of the small furs before the annual supply ship sailed.

A century after Cockqua's white visitors departed, linguist John P. Harrington ventured to the vicinity of the same village in search of elders who could still speak their tribe's vanishing language. When he asked about *sewelel*, a woman named Emma Luscier responded immediately. "My mother had one of those robes," Luscier said. "They were high-priced blankets. The skins were treated with oil so they were soft as a cloth & were sewed together so neatly that you could not see the seam. And we eat the meat. That *sewelel* eats only 2 kinds of grass: sour clover and a sedge about 2 ft. high that cuts your hand." While David Douglas did manage to send his 27-hide mountain beaver robe back to England, he never caught a glimpse of a live *sewelel* or understood quite how they fit into the culture of Cockqua's people.

VIII.
CROWN OF THE CONTINENT
SPRING–SUMMER 1827

A Thousand Miles Upstream

On March 20, 1827, as the spring express bound for Hudson Bay prepared to depart from Fort Vancouver, David Douglas waited beside the landing with a small kit, a handcrafted tin box filled with seeds, and mixed feelings: "Though I hailed the prospect of returning to my native land, I confess that I could not quit such an interesting country without regret." He took some consolation in the farewells of George Barnston and other members of his "little society," and in the fact that Dr. McLoughlin, Edward Ermatinger, and Alexander McLeod would be along for the trip up the Columbia.

The "gentlemen" passengers (agents, clerks, and visiting naturalist) alternated between riding in the canoe and walking on shore around the familiar portages of the Cascades, the Dalles, and Celilo Falls. Reveling in clouds of spring butterflies and migrating swallows, the collector plucked one small mustard new to his acquaintance. By March 26, they had negotiated all the rapids without mishap and were lunching on the upper end of Blalock Island when Dr. McLoughlin realized he had left his gun at their breakfast stop. Douglas could understand how the chief factor would be "loath to lose it, having some celebrity attached to it (Sir Alexander Mackenzie used it on both his former journeys)."

While three men hurried back to fetch the relic, Douglas watched a group of Umatillas baking cactus in underground ovens in much the same manner that camas roots were prepared, but if he tasted the results, he did not record his reaction. The voyageurs clearly preferred horsemeat, which they purchased from the Indians for supper that evening. After taking his turn at watch and "cooking by the kettle" with Alexander McLeod, Douglas set off on foot early the next morning in hopes of finding some new dryland blossoms he might have missed on his previous trips. Instead he bagged a young "grouse of the plain" (sage grouse). Although he didn't fancy the bird's adolescent plumage, during a quick dissection he was impressed with the smoothness of its gizzard and the sheer size of its windpipe—"fully stronger than a goose."

Over the next few days, as the express moved through the heart of the shrub-steppe, clouds of these birds often rose from shore and were sometimes seen dancing, "most likely holding their weddings." Camped below Priest Rapids, Douglas rose before dawn to crouch among the sagebrush and observe a sage grouse lek in its full spring frenzy.

> The males spread the tail like a fan and puff up their breast or pouches to as large in size as the whole body, and like the pigeon singing their song, which I listened to with much pleasure. Their voice is "hurr-r-r-r hoo hurr-r-r-r hoo," a very hollow, deep melancholy sound. The female I have heard call only when rising from the ground, which is "Cack-cack-cack" like the common pheasant.

Always bent on acquiring a matched pair of any animal or bird, he shot a hen, skinned it, and wrapped it in his bundle. He and McLeod brought down two cocks each, all of them too damaged to preserve but plenty good to eat.

Upstream from Priest Rapids, the river terraces were still covered with snow. Lacking fresh plants to claim his attention, Douglas whacked mineral samples from nearby cliffs with his new hammer. Approaching Fort Okanagan on April 5, he was heartened to see Archie McDonald striding down from the post to meet the express. The friends strolled along the banks, exchanging news and "picking up any mineral that seems curious; found some very fine pebbles." Long days of walking on rocky shores in deerskin moccasins had rendered Douglas's feet "painful, blistered, and blood-run," and when the canoe re-launched from Okanagan, the tender-footed collector climbed aboard to ride for awhile. As the voyageurs worked their way up Nespelem Canyon, he marveled at enormous driftwood pines stranded on rocks forty feet above the present water level—testament to the power of spring runoff.

On the evening of April 7, the canoe party made camp at a landmark known as Grosse Roche, or Big Stone. Meanwhile, Ermatinger, McLoughlin, and McLeod, traveling overland on horseback to meet them, mistook the rendezvous spot and spent several hours waiting for the boat before realizing their mistake and trotting into the camp at Big Stone, where they immediately roused Douglas from a sound sleep to join them in a late supper. For once "the good jokes at losing the way" were not at his expense.

By the time the express stopped at Fort Colvile on April 12, the gentlemen were laden with curlews and grouse they had shot along the way. John Dease was happy to see his former guest again, and John Work proudly produced a fine pair of mountain goat skins he had purchased from Indian hunters. Douglas would have liked to take the goats back home to England, and he spent a few moments coveting them before learning they were destined for Nicholas Garry, deputy governor of the Hudson's Bay Company. "I of course could not ask for them. Mr. Sabine, I hope, will get them through that channel." Work was also preparing a coyote hide, an animal that Douglas thought "curious from its being the deity or god of the Flathead tribe of Indians." Coyote, or *spill yay*, did not quite correspond to the Presbyterian notion of a deity, but did figure prominently in Salish folklore as a supernatural trickster and benefactor.

At about the same time that the collector was admiring Work's coyote pelt, several live canines who hung about the fort were sniffing around his tent. Wary of varmints, Douglas had wrapped his two sage grouse skins in a small oilcloth and hung them from the tent poles with a leather thong, but somehow the dogs reached the birds and tore them to pieces. He found himself "grieved at this beyond measure. Carried the cock bird 457, and the hen 304 miles on my back, and then unfortunately lost them." He was far from giving up on such a prize, however, and the next morning wrote to Archie McDonald asking him to shoot another pair of grouse and get them aboard the next vessel bound for England.

Thinking of other items that remained on his wish list, Douglas composed a reminder "to be read to my Chenook friend Cockqua" concerning those elusive *sewelel*. For John McLoughlin, who was returning to Fort Vancouver, he penned instructions regarding two boxes of specimens and personal belongings he had left there to be loaded on the next ship. Then the collector gathered bulbs from some of his favorite early spring flowers—yellow bells, glacier lilies, and spring beauties—to take with him on the overland journey, remarking that although the season was not ideal for transplants, "I cannot forbear making a trial."

After an evening horseback ride with McLeod and McLoughlin and a farewell dinner at the post, he said goodbye to "my Columbian friends . . . a grateful remembrance of which I shall ever cherish." In order to expedite their morning departure, Douglas, Edward Ermatinger, and their seven-man crew spent the night of April 17 at the boat launch a short distance above Kettle Falls. Contrary to their intentions, "we overslept ourselves this morning and were not up until daylight, when we hurriedly pushed off lest we should be seen by our old friends."

As the chagrined paddlers bent to their work, Douglas took in the novel landscape of the upper Columbia. His journal entries doubled in length, and the mountainous terrain inspired new heights of lyrical prose.

Sky beautiful at sunset, the snowy summits of the hills tinted with gold; the parts secluded from his rays are clothed with cloudy branches of the pine wearing a darker hue, while the river at the base is stealing silently along in silvery brightness of dashes through the dark recess of a rocky Dalle. How glad should I feel if I could do justice to my pencil (when you get home, begin to learn).

Upstream from Kettle Falls, new varieties of flora and fauna appeared along the river. Douglas noted an abundance of hazelnut bushes and huge birch trees around Sheep Creek near modern Northport, Washington. He compared tall, slender western white pine with the sugar pines in the Umpqua, and watched a chickadee, with "a sweet chirping voice . . . hang by the claws, head down, from the cones of the pine." At the dramatic confluence of the Pend Oreille River on the forty-ninth parallel, he reflected that Captain William Clark had skirted the upper reaches of this same drainage in 1806.

On April 20, the party passed the mouth of the Kootenay River (near present-day Castlegar, British Columbia) and entered Lower Arrow Lake, where "high snowy peaks are seen in all directions raising their heads to the clouds." When a light breeze sprang up, the voyageurs hoisted a sail and began to sing their way up the lake, but Douglas, who for some reason had turned to linguistic work, was in no mood for a chorus. "Intended to have arranged a few words of the Chenook language," he fretted, "but was molested out of my life by the men singing their boat-songs."

Over the next few days, as they made their way through the Lower and Upper Arrow Lakes, the men stopped at several villages of the Lakes (Sinixt) tribe to trade for whitefish, suckers, and wild game. From the variety of hides around their lodges, Douglas judged that the Lakes people lived very comfortably. Many of the skins belonged to woodland caribou (which he called reindeer), and the naturalist considered how the bucket-sized feet of the animal served as "a proof of the wise economy of Nature, given it to facilitate its tedious wanderings in the deep snow."

While Ermatinger bartered for *pas d'ours* (bearpaw snowshoes) at one encampment, Douglas watched a group of Lakes people gather black lichen from the trees, and examined one of their sturgeon-nosed canoes that he had been admiring on the river.

They are 10 to 14 feet long, terminating at both ends sharply and are bent inwards so much at the mouth that a man of middle size has some difficulty in placing himself in them. One that will carry six persons and their provisions may be carried on the shoulder with little trouble.

The canoe was fashioned from several tree species: cedar and willow for the frame; a single piece of western white pine bark for sheathing; strips of birchbark to skirt the gunwales; the bark of cedar and chokecherry to tie together some of the elements, and the elastic roots of Engelmann spruce for others. Pine pitch mixed with deer tallow gummed all the seams.

When the furmen stopped at a Lakes camp to purchase bear meat for supper one evening, Douglas traded for a quantity of mountain goat wool in exchange for seven musket balls and powder charges. Back at Fort Colvile, John Work had given him a nightcap woven by an Indian girl from goat's fleece; impressed with its warmth, Douglas envisioned similar comfort for his feet. "Get a pair of stockings made of it," he jotted in his journal.

Douglas characterized Edward Ermatinger, two years his senior, as "a most agreeable young man." He admired the trader's skill on a flute, but thought him lax at keeping the Sabbath. "A Sunday in Great Britain is spent differently from what I have had in my power to do," he mused. "Day after day without any observance passes, but not one passes without thoughts of home." On the same Sunday that he decried Ermatinger's unrelenting work habits, Douglas himself engaged in commercial transactions, worked at his botanical trade, and spent far more time in contemplation of the scenery than of scripture.

After setting off at their usual hour of 4 a.m. on Monday, April 23, the express approached a sharp bend where the river "to all appearances loses itself in the mountains." The current picked up speed, and rapids became more frequent. Two days later the men halted for breakfast at the foot of the formidable Dalles des Morts, or "Narrows of Death, a terrific place in the river, which takes its name from a tragical circumstance . . . when ten individuals endured almost parallel sufferings, and were finally all released by death, with the exception of one." While the crew set about portaging the baggage and hauling their bateau up the rapid, Douglas was overcome by "a melancholy sensation of no ordinary description, filling the mind with awe on beholding this picture of gloomy wildness."

Awe overpowered melancholy over the next couple of days. On the morning of April 27, as the canoe rounded a short turn in the river, "one of the most magnificent prospects in Nature opened to our view." Within a few miles, the paddlers reached the Columbia's northernmost turn at the mouth of the Canoe River and beached the bateau on a low point called Boat Encampment, the site of David Thompson's canoe-building exploits sixteen years before. From a spot now drowned beneath the backup from Mica Dam, Douglas gaped at the spectacular and intimidating mountains that he would soon attempt to cross on foot:

> On beholding the Grand "Dividing Ridge" of this mighty continent, all that we have seen before seems to fade from the mind, and to be forgotten in the contemplation of their height and indescribably rugged and sharp peaks, with the darkness of the rocks, their glaciers and eternal snows.

Later he struggled to describe "a feeling beyond what I can express. I would say a feeling of horror."

As soon as the boat was unloaded, he examined the seeds in his tin box; relieved to find them dry and undamaged, he turned his attention to his wardrobe, which consisted of four shirts (two linen and two flannel); three handkerchiefs; two pairs of stockings; two sets of jacket, vest, and trousers; and seven pairs of deerskin moccasins. In preparation for the trip across the Divide, Ermatinger loaned him a pair of wool leggings fashioned from the sleeves of a voyageur's old capot.

Looking around the campsite, Douglas found a single lily just emerging, as well as the same species of twinflower he had seen at the great river's mouth, fully one thousand miles downstream. Perhaps that small flower reminded him of all the territory he had covered in the two years since he had first noticed its nodding pink blossoms, for in his journal that evening he cataloged his "wanderings on the Columbia and through the various parts west of the Rocky Mountains." After listing the estimated distances for his many expeditions and side trips, he calculated that he had traveled an astonishing seven thousand and thirty-two miles. Measured on a modern map, that seems like a pretty fair estimate.

That evening, Ermatinger celebrated the successful completion of the journey to Boat Encampment with a supper of fresh goose and the last of the potatoes he had brought from Fort Colvile, remarking that these tubers were "probably the first ever eaten at this place." Douglas was grateful for the extra starch the next day, as he struggled to learn the rudiments of walking on snowshoes. Carrying his tin box and oilcloth-wrapped journals on his back, he watched dun-colored snow buntings flit past and gazed at blue Steller's jays along the mountain slopes. That evening, his shoulders aching from his heavy load, he took a shot at a wolverine that wandered into camp, but the animal escaped, depriving the collector of another curiosity to take home.

Another Hemisphere

The party crested Athabasca Pass on May 1, and while the crew rested, Douglas made an ambitious trek up Mount Brown. Sharp pains in his ankles and knees awoke him before dawn the next morning, but he had crossed the Continental Divide, and a little stiffness was not going to stop him now. Pausing for a rest at the traditional fur-trade toasting spot known as the Committee's Punch Bowl, he viewed one more spectacular glacier with "columns and pillars of ice running out in all the ramifications of the Corinthian order." After that the snow depth rapidly diminished as the party descended, and by midmorning he was hanging his cursed bearpaws on a tree for the use of a future traveler.

Striding down the east slope of the Rocky Mountains with gray jays fluttering about his head, he found himself reinvigorated. By midafternoon the temperature had climbed to an oppressive 57 degrees, and he strolled from gravel bars covered with dryas flowers into woods sprinkled with wild sarsaparilla. It was early evening before he realized that he had, once again, wandered far away from his comrades. Just before the sun dipped behind the mountains, he spied a wisp of smoke about a mile to the east and quickly made his way toward it.

Jacques Cardinal, a voyageur who was on his way to meet the express with a string of fresh horses, welcomed Douglas into his camp and shared a roasted shoulder of bighorn sheep. Cardinal had no spirits to offer, but for liquid refreshment he pointed to the nearby stream. "This is my barrel," he laughed, "and it is always running." As Douglas tucked into his meal, Cardinal showered him with the latest news of the Franklin expedition, whose scattered members were also on their way back to England. Dr. John Richardson was far downstream on the Saskatchewan. Much closer by, another of Franklin's crew, an "old botanical acquaintance" of Douglas named Thomas Drummond had spent the summer botanizing up the Athabasca River and into the Rockies, traversing the same trail that Douglas

had just snowshoed. For the first time since he came to the
Northwest, there was the chance that he might come face
to face with other professional collectors.

After Douglas, Cardinal, and the pack horses rendezvoused with Ermatinger the next morning, the collector
resumed his wandering ways, tramping through the sodden
valley of the Athabasca. He fired at spruce grouse and found
a nice blue anemone in bloom, all the while marveling at the
unfamiliar ecosystem:

*The difference of climate and soil, with the amazing disparity in
the variety and stature of the vegetation, is truly astonishing, one
would suppose it was another hemisphere, the change is so sudden
and so great.*

When they reached the first navigable stretch of the
Athabasca River, Douglas wrestled the horns and partial skull
of a bighorn sheep he had bartered from Cardinal into a birchbark canoe with Ermatinger and the crew. They swept down
to the small establishment at Jasper House to dine on whitefish, then listened to one of the voyageurs scratch out dance
tunes on an old violin. The next morning, rogue cakes of ice
slowed their progress, and the collector's initial thrill at water
travel through the northern spruce forests began to wane.

May 5th. This day admits of little variety.

*May 6th. Country the same as yesterday. . . . Burnt my blanket
and great toe at the fire last night.*

*May 7th. The whole distance of this river . . . admits of no variety;
seeing one mile gives an idea of the whole.*

He barely allowed his great toe to touch the floor at Fort
Assiniboine before he joined a party heading out to meet

canoes coming down from Little Slave Lake under the command of John Stuart, who had handled the surveying duties during Simon Fraser's run down the Fraser to the Pacific two decades before. Over a supper of caribou steaks, Douglas found that Stuart had an intimate knowledge of the surrounding country, as well as "a good idea of plants and other departments in natural history." One department in which Stuart specialized was spruce grouse, which he called "White Fleshers" because of their delicate meat. As they traveled back to Fort Assiniboine, Douglas complained about "this uninteresting wretched country affording me no plants," but he and Stuart did bring down some fine white fleshers.

From Fort Assiniboine, an overland trail led to Fort Edmonton on the North Saskatchewan River. Douglas became "most anxious to learn the fate of my packet of seeds"—a box of precious specimens and seeds that Finan McDonald had carried across the Divide the previous fall and was holding for him at Edmonton. Impatient to find out its condition, the collector set off afoot on May 21, guided by an elderly Nipissing Indian who proved to be an excellent walker despite his advanced age. With Douglas's small hatchet, the pair threw down trees to bridge streams and creeks, then built a raft to cross the swollen Sturgeon River. They were slogging across the spring-flooded prairie at dusk when they heard the "sweet music" of camp dogs and knew they must be nearing the post. Even though night was closing in and neither he nor his guide had eaten since 4 a.m., Douglas paused to make himself presentable.

Being all over with mud, I returned half a mile to a small lake, stripped and plunged myself in and then comforted myself with a clean shirt which I carried on my back in a bundle.

The chief factor at Fort Edmonton cheerfully received his well-scrubbed guest and had the cook prepare moose steaks for his dinner. The festive atmosphere was dampened when Finan McDonald informed Douglas that his specimen box had been injured during the journey over the mountains. Douglas decided to wait till morning to assess the damage, but fretted when he learned that the post factor had asked naturalist Thomas Drummond to open the box and change some of the papers. "This was kind," the protective Douglas wrote, while admitting that he did "not relish a botanist coming in contact with another's gleanings."

Overwrought with fatigue and worry, the collector passed a sleepless night, then rose at first light to view the contents of his box. To his great relief, he found the seeds in much better order than he expected. Of the pressed plants, only eighteen papers had been significantly damaged. Unfortunately, one of those happened to be the Brown's peony. "This, one of the finest plants in the collection. It often happens that the best goes first," he lamented. He then wrote one of his little reminders to himself that in the future, specimens should be stored in a soldered tin box, then packed inside a strong wooden box for shipping.

While the post carpenter made two such wooden boxes to transport the collector's belongings and new accumulations downstream, Douglas spent time studying a tame golden eagle that had been captured as a chick and raised at Fort Edmonton. He had heard a great deal of lore about the strength and ferocity of the bird the voyageurs called the calumet eagle, in reference to the ceremonial pipestems that were often decorated with its tail feathers. There was no such tail for the naturalist to examine on this bird, however, for some of the boys who lived at the fort had recently plucked it clean. The factor suggested that Douglas take the bedraggled eagle to England, and so with some difficulty he maneuvered the sharp-taloned bird into a cage.

By May 25, the traders who had rendezvoused at Fort Edmonton were ready to proceed downriver, and in celebration of their imminent departure, Ermatinger indulged the group with a tune on his violin. "No time was lost in forming a dance," Douglas wrote, "and as I was given to understand it was principally on my account, I could not do less than endeavor to please by jumping, for dance I could not."

The next morning, with sheep horns, caged eagle, and two new wooden boxes in tow, Douglas accepted an invitation to ride in John Stuart's canoe, where they could continue their conversations on natural history. As the brigade moved past moose, pronghorn antelope, coyotes, and two swimming elk during the next week, Douglas grew increasing irked with the scant opportunities to walk on shore.

When a buffalo herd was spied within range on June 2, the paddlers put ashore, excited at the prospect of fresh meat. While the hunters fanned out in pursuit, the collector stretched his legs and happily botanized along the banks. Late that afternoon, hearing cries of alarm, he hurried to the scene to find an enraged bison mauling Finan McDonald, who had been tracking the wounded bull when it turned and "rushed toward him with the utmost impetuosity." Douglas and several other members of the party witnessed Finan's savvy reflexes:

> Seeing that it was utterly impossible to escape, Mr. McDonald had the presence of mind to throw himself on his belly flat on the ground, but this did not save him. He received the first stroke on the back of the right thigh, and pitched in the air several yards. The

wound sustained was a dreadful laceration literally laying open the whole back part of the thigh to the bone; he received six more blows, at each of which he went senseless. Perceiving the beast preparing to strike him a seventh, he laid hold of his wig (his own words) and hung on; man and bull sank the same instant.

The onlookers stood paralyzed in horror as their comrade, clasping desperately at the bull's cape, disappeared into the tall grass and fading light of the summer evening.

Poor McDonald was thus situated for two hours and a half, bleeding at the point of death, and that too under cloud of night, which afforded us scarcely any opportunity of rescuing him, for the animal lay watching within a few yards, and we were afraid to fire, lest a shot should strike our friend.

Finally, an accidental round fired by one of the hunters spooked the buffalo. The animal raised itself up, "first sniffing his victim, turning him gently over, and walking off." The men rushed to the unconscious McDonald to find that the bull's horn had struck him directly on the left side, and

had it not been for a strong double sealskin shot-pouch, with ball, shot, wadding, &c, which shielded the stroke, unquestionably he must by that alone have been deprived of life, being opposite the heart. The horn went through the pouch, coat, vest, flannel, and cotton shirts, and bruised the skin and broke two ribs. He was bruised all over, but no part materially cut except the thigh.

Like many travelers of the time, Douglas always carried a lancet in his pocket. Concerned that McDonald had not lost enough blood, he bled the battered fellow some more, then "bound up his wounds, and gave him all the aid that a small medicine chest and my slender knowledge of surgery would suggest." Twenty-five drops of laudanum produced a deep and effective sleep, and McDonald was loaded into a canoe that traveled through the night to Carlton House, where he could receive the medical attention of Dr. John Richardson. But the physician had moved on to Cumberland House, the next major fort on the Saskatchewan. "In order that Mr. McDonald may get surgical aid as soon as possible," Ermatinger immediately dispatched a fast canoe to carry the injured man downstream.

Douglas remained at Carlton House, where Thomas Drummond was also in residence. During the course of the first evening, the two collectors compared experiences. "He appears to have done well," Douglas observed. "I must state that he liberally showed me a few of the plants in his possession—birds, animals, &c. in the most unreserved manner." Douglas had some extra weeks to continue his botanical pursuits before meeting the ship at Hudson Bay, but after touring several interesting habitats around Carlton House with Drummond, he decided that to remain in that vicinity "might be looked on as an encroachment by him [Drummond]," and resolved to look farther afield.

Downstream at Cumberland House on June 10, he finally crossed paths with John Richardson, who had amassed a "splendid herbarium and superior collection in almost every department of natural history," and encouraged Douglas's idea of botanizing around the Red River. As for Finan McDonald, he had been taken under Richardson's attentive care. Apparently the good doctor did not think much of his patient's chances, because in a book published after his return to London, he reported that McDonald had succumbed to the buffalo's stomps. But Richardson was mistaken: McDonald recovered from his injuries and retired to Glengarry County, Ontario, where he thrived for another twenty-five years.

Red River

Douglas declared the stretch from Cumberland to Norway House on the northeast tip of Lake Winnipeg "no place for botanising," but he did shoot a fine white pelican, and even though some mischievous boys damaged the bird's neck by playing with it, he managed to preserve its skin. At Norway House he intercepted a packet of letters that had come by way of Hudson Bay. Among them were pleasant tidings from Sabine and Hooker, but "news of a melancholy cast" from his brother John: their father had passed away at Scone.

Douglas could not dwell on the loss, however, because the next morning (June 17) Governor George Simpson arrived on one of his whirlwind tours. Although effusive in his attentions to the naturalist, Simpson was slightly distressed at the ragged state of his clothing. "He offered me some linen and a coat, which I refused," Douglas remarked. Soon after the governor departed, Sir John Franklin stopped at Norway House on his return from the Arctic. When he learned that Douglas intended to go to Red River, the explorer kindly offered him a seat in his canoe, traveling south on the lake. By this time Douglas had adopted a live bald eagle to complement his golden one, but there was no room for them in the boat. Thrilled with the chance to ride with one of his heroes, he shipped the calumet eagle to Hudson Bay, left the white-headed one with a woman "attached to the establishment," and hopped on board.

John Franklin practiced a different mode of travel than the fur brigade, stopping frequently to investigate on land, and Douglas piled up more plants as they coasted along Lake Winnipeg's eastern shore: grasses and sedges, chickweeds and campions, the first oak he had seen in months, and an Indian hemp not nearly so strong as the one he knew from the Columbia. When they reached Fort Alexander at the mouth of the Winnipeg River on July 8, Douglas bade farewell to Franklin, who turned east toward Montreal.

Finan McDonald's brother John was the agent in charge of Fort Alexander, and he quickly found a guide who agreed to carry Douglas to the Red River settlement for the price of four dollars. The voyageur steered them beneath vast flocks of passenger pigeons and through hordes of mosquitoes for three days until on July 12 they reached the burgeoning Red River colony, a sanctuary for Scottish immigrants and mixed-blood fur trade families. Douglas, who had not seen a town in three years, marveled at the "thinly planted low houses, with small herds of cattle wandering from the fold. Humble and peasant-like as these may appear to many, to me . . . they impart a pleasant sensation." His pleasure was returned in kind as he walked along a path into the village.

> Strangers in their quarter appear to be few: scarcely a house I passed without an invitation to enter, more particularly from the Scottish settlers, who no doubt judging from my coat (being clothed in the Stewart or royal tartan) imagined me a son from the bleak dreary mountains of Scotland, and I had many questions put to me regarding the country, which now they only see through ideal recollection.

Passing a school, he stopped to chat with two young boys. When they assured him they were avid readers of parables and popular storybooks, he handed them "a few trifling articles," then watched them run off to display the windfall to their schoolmates. Nearby he saw the great windmill on the Assiniboine River and the stockade of Fort Garry, named for the Bay Company governor who had sponsored his Northwest adventures. Douglas walked through the gates and introduced himself to chief factor Donald McKenzie.

The factor placed a large tureen of milk on the table for his visitor and began to talk. "His conversation to me is the more acceptable from the intimate knowledge he possesses of the country west of the Rocky Mountains," Douglas remarked, well aware that McKenzie had traveled overland to the Columbia in 1811, shepherding Philadelphia naturalist Thomas Nuttall along the Missouri for the first part of the journey. McKenzie had also pioneered the first expeditions to the Snake River and established Fort Nez Perce on the Walla Walla in 1818. Douglas was probably thinking of his own discomforts when he remarked that McKenzie had, "like all who share in such undertakings, shared in the fatigues and hardships attendant on these expeditions."

Later that afternoon, Douglas received a surprise visit from an Indian lad who attended the Protestant school. Known to the missionaries as Spokane Garry, he was the son of Ilum-Spokanee, whom Douglas had met at Spokane House. Two years before, Governor Simpson had convinced the Spokane chief to send the boy to the Red River school, and Garry had become part of the first of many generations of Indian children removed from their homes in the name of a proper Christian education. Upon learning of Douglas's arrival from the Columbia country, Garry had come "to inquire of his father and brothers, whom I saw." Discovering that he spoke good English, Douglas relayed news of Garry's family and his faraway home. During their conversation, Garry confided that he had almost forgotten his mother tongue, a lament that would be repeated by tribal boarding school students for the next one hundred and fifty years.

On Sunday, Douglas set out to attend the Presbyterian church in the settlement, but lost his way. A boy was dispatched to find the wandering naturalist, and brought him in halfway through the service. Douglas wrote that the incident reminded him of "the man of the world who, in the parable, was compelled to go to the feast by the person stationed on the wayside."

There was also a Catholic Church in the village, led by Bishop Provenchier, "a very tall stout man who spoke English with that broken accent peculiar to foreigners . . . but appears to be a man of most profound acquirements." Over the next three weeks Douglas spent considerable time with the bishop, soaking in his wealth of local knowledge. Provenchier explained the métis culture that was growing up around the Red River colony, and their "domestic economy, farming, spinning, and weaving cloth from the wool of the buffalo." He recounted the devastating floods of the previous year, which had almost wiped out the whole establishment. Douglas quizzed the bishop for plant information—learning, among other tidbits, that the native "hops do very well for beer"—and, as his stay drew to a close, he stopped by the Catholic church for Sunday services.

The collector's 1827 plant list had reached #291 when he asked Donald McKenzie to arrange for a boat to carry him back to Norway House. When he departed on August 10, his knapsack was well stocked with provisions from McKenzie and mail from fur traders and Scottish colonists to friends and relatives in Great Britain. As he made his way out of town, many settlers stepped out of their cottages to present him with small cheeses as tokens of their appreciation for his postal services.

Blown Across the Bay

Douglas arrived back at Norway House to find his roots, bulbs, and bird skins in good order, but his bald eagle had died of starvation because it refused to eat fish. On August 18 Douglas embarked on the home stretch of his transcontinental journey, descending the Nelson River toward York Factory in a flurry of white water. The collector was pleased to breakfast near a spot known to the voyageurs as the Tea Islands, where an abundance of Labrador tea grew. He didn't seem too alarmed when the boat struck a rock while shooting a rapid, shattering seven planks and splashing water over his boxes. "Just had time to reach a small island when she was filled," he wrote. "My hands tied up—could not get off." While the boat was being repaired, he collected wild tea plants, chiding himself for not digging some live Columbia River salal to carry to London: "Why did you not bring *Gaultheria* alive—across the continent—2900 miles? It could be done."

Floating beneath brilliant northern lights, the canoe reached York Factory on the western shore of Hudson Bay at the end of August. The chief factor there, perhaps with a nudge from George Simpson, had ordered the post's tailor to sew up a new suit and linen for Douglas's homeward voyage. The only bad news concerned his golden eagle, which had strangled itself on its tether cord. "What can give one more pain?" moaned Douglas. "This animal I carried 2000 miles and now lost him, I might say, at home."

Three members of the Franklin Expedition—Captain George Back, Lieutenant E. N. Kendall, and Thomas Drummond—were all scheduled to sail for London with Douglas, and as the passengers made their final preparations at York Factory, he squeezed one last line onto the final page of his notebook: "It now only remains to state that I have great assistance, civility, and friendly attentions from the various persons I have formed an acquaintance with during my stay in North America."

On September 1, a skiff powered by eight oarsmen ferried Douglas, Back, Drummond, and Kendall into the shallow bay to dine with the captain of the *Prince of Wales*, which lay at anchor half a dozen miles offshore. During their return, a fresh breeze suddenly rose into a gale. After an unsuccessful attempt to return to the *Prince of Wales*, Captain Back took charge, ordering the oarsmen to set an anchor and ride out the storm. But the anchor tore loose, and howling winds dismasted the skiff, then pushed it out into the great bay. The small boat was laden with casks of provisions; Back had these jettisoned and attempted to gain some headway by rowing, but to no avail. With every man aboard seasick from the frightful chop, there was nothing to do but bail for all they were worth to keep the little craft from sinking.

Captain Back stayed at the tiller while Kendall and some of the men erected a temporary mast and hoisted a small sail. After a debate about whether to drive the skiff onto the rocky shore, Back decided to let her run before the wind. Lightning flashed and violent winds continued to blow through the night, and only Back's skill at the helm kept them from being caught between the crashing waves. Everyone on board remained "in so benumbed a state, that it was hardly possible to make the necessary exertion to keep the boat from sinking."

The storm raged well into the next day, and even after the wind abated, the sea remained too rough for the men to make any progress with the oars until the second night. By then they had been blown more than sixty miles from the fort, and Douglas was "dreadfully ill." Drummond later recalled cracking open a cask of oatmeal, "which mixed with a little salt water, sufficed to allay our hunger; but I believe that Lieutenant Kendall and myself were the only partakers."

The *Prince of Wales* sent out a dory with more relief, and the skiff slowly made its way back to York Factory.

The ship finally weighed anchor a few days later and enjoyed a quick four-week cruise to Portsmouth, but the effects of the ordeal in the skiff lingered. Drummond wrote that "Mr. Kendall and Mr. Douglas suffered severely, and did not recover the full use of their limbs until their landing in England." William Hooker later remarked that Douglas was confined to his bed for most of the voyage. The naturalist, who described the rest of his New World adventures so vividly, apparently never wrote a single line about his storm-tossed night on Hudson Bay.

IX.
"A Scientifick Naturalist"
Fall 1827–Fall 1829

One Pine, Four Lilies

After disembarking from the *Prince of Wales* in Portsmouth in early October, Douglas made his way to London, where he delivered dispatches to the Hudson's Bay Company headquarters. He would have called at the Horticultural Society's office on Regent Street, then continued five miles west to the Chiswick gardens. His friend William Booth, still employed as a clerk by the Society, had the honor of greeting the weather-beaten traveler.

> *His appearance one morning in the autumn of 1827 was hailed by no one with more delight than myself, who chanced to be among the first to welcome him on his arrival.*

As heartening as this reunion must have been, Douglas would have found a tour of the grounds equally gratifying, for many of the seeds he had shipped from the Columbia two years before were now flourishing in the gardens.

Unpacking his latest treasures, he paid particular attention to the two tin boxes he had so carefully tended across the continent, and he applied his literary abilities to one of the species whose seeds rested within. A mere three weeks after his return, he wrote to Joseph Sabine:

> Understanding upon my return from North-west America, that considerable interest has been excited by reports of a new species of Pinus of gigantic size having been discovered by me, I beg permission through you to lay a short account of it before the Linnean Society.

A few days later, on November 6, 1827, Sabine read Douglas's description of the sugar pine to the monthly meeting of the Linnean Society, England's preeminent natural history organization of the time. Whether its author was still recovering from the rigors of his ordeal on Hudson Bay or was too shy to take the podium before some of the most influential scientists of the day is not known, but the audience would have found nothing unusual in Sabine's delivery—members often read letters from contributors at their meetings.

The annual edition of the Linnean Society's *Transactions* included a transcript of Douglas's account. In clear, concise prose, he portrayed the princely tree of the Umpqua Mountains, noting the sandy soil and low hills where it attained the peak of its growth. He recounted his measurements of one fallen giant and analyzed its bark, branches, crown, needles, and cones. He described the membranous wing that surrounded each seed as having "an innumerable quantity of minute sinuous vessels filled with a crimson substance, and forming a most beautiful microscopic object."

He also addressed the tree's economic possibilities, principally its fine white wood. He had brought a sample of the amber-colored resin home with him, and measured its specific gravity at 0.463. A touch of fire, he noted, imparted a distinct sugary taste to the sap, which the Umpqua Indians mixed with their food as a sweetener. They also gathered and roasted the seeds of the tree, which they called *na't-cleh*, eating some of them whole and pounding the rest into coarse cakes for winter fare. He believed that the tree's range extended into northern California, for Franciscan monks in Monterey had served Archibald Menzies a dessert of large pine seeds during his 1793 journey, which he and Douglas agreed must have come from this same species.

Douglas's discovery of this new pine had earned him the naming rights, and he chose to honor Aylmer Bourke Lambert, author of the standard monograph on conifers and the sitting vice president of the Linnean Society. After announcing this tribute, the collector summarized the distinctive character of *Pinus lambertiana* with thirteen carefully chosen Latin descriptors that ushered the new species into the annals of European science.

One factor regarding the oversized pine that Douglas could not address was its viability in the English climate, but the seeds extracted from the three cones he had brought home would eventually provide the answer. The gardeners at Chiswick planted as many as possible, then distributed additional seeds to Britain's principal nurserymen. Several of the Society's fellows eagerly volunteered to raise sprouts in their own gardens, but none of the seedlings met with any lasting success.

Among naturalists of all persuasions, Douglas's company was as eagerly sought as his seeds. He dined with fellows of the Linnean Society and attended a social salon presided over by Joseph Sabine's sister and her astronomy-loving husband. William Booth, with whom the collector shared lodgings at Chiswick, later wrote:

His company was now courted, and . . . he could not withstand
the temptation (so natural to the human heart) of appearing as one
of the Lions among the learned and scientific men in London.

Douglas apparently relished entertaining these gatherings
with tales of running from hostile Indians, eating dried animal skins when he ran short of food, and training his "botanical" horse to stoop when passing beneath tree branches so as
not to dislodge the plant bundles it carried on its back.

According to Booth, Douglas basked in his new popu-
larity: "Flattered by their attention, and by the notoriety of
his botanical discoveries, he seemed for a time as if he had
obtained the summit of his ambition." And by all outward
appearances, Douglas was indeed thriving. Many of his botanical discoveries were displayed at meetings of the Horticultural
Society, and their *Transactions* featured accounts of selected
Columbia plants, illustrated by large colored engravings. John
Lindley, assistant secretary of the Society, edited a periodical
called *Edwards's Botanical Register* and in several successive
issues he highlighted Douglas's work. Lindley chose species that were growing successfully in the Chiswick gardens,
enriching his descriptions with Douglas's comments on their
native habitats. The accompanying colored plates depicted
each flower as a sumptuous work of art.

The wide coverage of his successes introduced Douglas's
expertise to the gentry of the scientific world, and he soon
ventured back into the halls of science, reading a paper on
mariposa lilies to a meeting of the Horticultural Society in
mid-February 1828. He reviewed the scant information on
the *Calochortus* group, based on a single dried plant belonging
to the species *elegans*, collected by Lewis and Clark. During
his explorations on the Columbia Plateau, Douglas believed

that he had detected three new species of "this interesting and highly ornamental Genus." The first, now commonly called sagebrush mariposa or green-banded star tulip, he found near the Dalles. After patiently coaxing several roots from the earth, he shipped them to England in 1826; transplanted at Chiswick, several had bloomed over the past summer. A white form whose beauty rivaled that of the sagebrush lily grew in higher mountain valleys. The third possible new species appeared on the dry grounds of the Columbia Basin between Priest Rapids and the Okanogan River. He had watched Plateau people gather its bulbs, which they called *koo-e-oop*, and had sampled some himself. "The root is roundish, crisp, and juicy," noted the discerning taster, "yielding a palatable farina when boiled." He had not been able to preserve a specimen, and the magnificent blue-flowered lily remained a tantalizing prospect for future botanists who might visit the Columbia.

Given the scientific and popular interest in the American West that existed in London, it probably came as no surprise when Joseph Sabine informed the Horticultural Society council on March 1, 1828, that publisher John Murray had offered to produce a narrative of Douglas's travels. The council members agreed to furnish "every assistance and facility . . . which the Society could command or bestow in aiding him therein," as well as granting him all profits from the publication. The most influential publisher in England, John Murray had brought Sir Walter Scott, Lord Byron, and many other prominent authors to print, and Douglas should have been thrilled at the prospect of appearing under such a prestigious imprint. But instead, his feelings were deeply hurt, according to William Booth, for Joseph Sabine had made all the arrangements with Murray without consulting Douglas. "This was done from the best of motive but by some means it failed of the desired effect," Booth observed. William Hooker added that both Sabine and Lindley offered Douglas their assistance, but "this proffered kindness was rejected by Mr. Douglas, as he had thoughts of preparing the Journal entirely by himself." Apparently the collector's independent streak was asserting itself, and he resolved to complete the task on his own terms.

As he began to shape his journals into a coherent account for a popular audience, the work of some of his most admired predecessors might have served as inspiration. William Bartram's romanticized narrative of his travels through Georgia and north Florida, for example, had sold out several editions in Europe in the 1790s. Thomas Nuttall's *Journal of Travels into the Arkansas Territory* combined technical botanical information with stories of exotic new places and hostile tribes. The narrative of Sir John Franklin's first Arctic expedition, published the year before by Murray, had stirred up a storm of interest in the polar regions and helped fund Franklin's second expedition. With similar experiences to offer, Douglas labored at his manuscript for many months, reworking his field journals into at least two drafts, but after a year had to admit that he was "still far behind." Hooker, who tried to encourage his protégé from afar, observed in frustration that Douglas "has much in his head but is totally unfit for authorship." The professor may have been too hasty in his judgment, for Douglas's two drafts displayed an engaging blend of personal observation and scientific awareness that captured the flavor of his Northwest adventures.

The Collaborator

Certainly there was much to fill Douglas's head at Chiswick as he applied himself on scientific and literary fronts alike.

He prepared a monograph titled *Some American Pines,* for example, that included lengthy treatments of seventeen conifers he had encountered in his journeys. Not surprisingly, he chose the tree now known as Douglas fir as the first entry. He characterized its unstalked cones as "ovate, pointed, pendulous in clusters at the extremities of the twigs, two to two and a half inches long." Each cone scale, soft and velvety to the touch, was fringed with marginal hairs and slightly notched at the base. Each distinctive bract glowed with a glossy reddish tint and extended beyond its scale about five-eighths of an inch.

Acknowledgment of his accomplishments was widespread. In early March, he was nominated as a fellow of the august Linnean Society.

> Mr. David Douglas, *residing at the Garden of the Horticultural Society, a gentleman well known from his travels, and the acquisitions which he has made in Natural History, particularly in Botany, on the continent of America, we, whose names are undersigned, do, from our personal knowledge recommend him as highly worthy to be elected a Fellow of the Linnean Society.*

The certificate of nomination was signed by such respected scientists as Robert Brown, A. B. Lambert, Archibald Menzies, and George Loddiges, the nurseryman to whom, three years before, Douglas had delivered seedlings from William Bartram's sourwood tree.

In addition to keeping an eye on his latest transplants from the Columbia River and shipping blossoming plants to Glasgow for Hooker to illustrate, Douglas worked assiduously at their classification, assisted at times by George Bentham, a friend of Hooker and Lindley. A recent arrival on the botanical scene, Bentham was dining at Lindley's home near the Chiswick gardens one evening in spring 1828, discussing the beautiful lupines Douglas had brought back from the Northwest, when the subject of their conversation stopped by. "He is quite a sauvage in his appearance and manners," the aristocratic Bentham wrote in his diary, but the two were soon poring over plants together. Employed as secretary to his uncle Jeremy, the influential Utilitarian philosopher, George had begun to apply his relative's rigorous logic to taxonomy, and he helped Douglas separate thorny Columbia species of monkeyflower, milkvetch, locoweed, and buckwheat. John Lindley, meanwhile, worked closely with the collector to classify his lupines and penstemons. In the *Botanical Register*, Lindley acknowledged Douglas for shedding new light on these "diverse and beautiful additions," several of which were soon hailed as commercial successes.

And yet all was not well in the garden. Joseph Sabine micromanaged every aspect of Horticultural Society business, and although Douglas revered the man, readjusting to life as an employee after three years of autonomy on the Columbia could not have been easy. A memo written by a Society employee on April 16, 1828, quoted Lindley as "quite disgusted with the manner in which he [Douglas] spoke to Mr. Sabine," and noted that "Mr. Douglas appeared to be quite ashamed of his conduct." This incident may have been precipitated when Douglas came upon a neglected box of bird and mammal skins that he had sent home in 1825, "expecting they would be a treat for Mr. Sabine." When Douglas pried off the lid, he found to his dismay that they had all been ruined by moths. According to William Booth, the box and its contents had been put aside by Sabine, who "would neither take the trouble to describe them himself nor allow them to be seen by others who were willing and competent to do so." That he had preserved the specimens with such great care and that they were now lost to science "mortified poor Douglas exceedingly."

Douglas seemed to be growing disenchanted with the social milieu as well. Booth wrote that "when the novelty of his situation had subsided, he began to perceive that he had been pursuing a shadow instead of a reality." By the summer of 1828, according to George Bentham, Douglas was dying to go out on expedition, but the Horticultural Society had no plans to send out any more collectors.

Douglas's frustrations did not, however, stem his natural curiosity and sociability with family and fellow naturalists. He had kept in touch with architect William Atkinson, whom he had first met as an apprentice at Scone Palace, and now spent many leisure hours at his home and gardens in St. John's Wood. When in London he sometimes stopped at the Hudson's Bay Company office to call on Nicholas Garry, who was always willing to converse about the Northwest. (Douglas may also have been keeping an eye out for the arrival of John Work's mountain goat skins). He attended meetings of the Linnean Society and dined with visiting naturalists from Europe and Asia. He compared specimens and notes with Robert Brown and Archibald Menzies, whom he fondly called the "good man of the mountain." In early June he and Menzies spent one whole day looking at grasses, "and just in the evening who came in but Scouler looking for me!" This was the first time that the friends had seen each other since parting at Fort Vancouver three years earlier, and the additional presence of the revered Menzies was especially gratifying. "Such a meeting!" Douglas wrote to Hooker. "The three North West Americans under the same roof." Scouler remained in London for the next two weeks, and besides reminiscing about their adventures, they discussed an account that he was preparing for the Zoological Society on the Chinookan practice of flattening the heads of their infants. When the article appeared, it was illustrated by a well-turned drawing of a Chinook cradle board, which Douglas called a "squeezing cradle."

> *I am indebted for the accompanying figure to my friend, Mr. David Douglas, who had equally with myself noticed this singular custom among the North-west American Indians.*

Douglas collaborated with other scientists as well. He presented the pair of condors he had preserved on the Columbia to the new museum of the Zoological Society, along with an account of their life history. He contributed another paper to the *Zoological Journal* describing the Pacific Northwest's deer and bighorn sheep. He offered his three hard-won grouse skins to Scottish artist James Wilson as models for a volume of zoological paintings. He donated his mineral collection to the Geological Society, and delivered cones of various conifers to Lambert for inclusion in a second edition of his landmark *Pinus*. He pored over his puzzling *sewelel* skin with John Richardson, who appended his thanks to the published account of the species: "I cannot avoid adding my tribute of praise to Mr. Douglas, for the . . . unusual liberality with which he has communicated his knowledge to the friends of science."

At the June 1828 meeting of the Linnean Society, dental surgeon and herpetologist Thomas Bell read a "Description of a new Species of *Agama*, brought from the Columbia River by Mr. Douglass." After describing the taxonomy and physical characteristics of the Columbia Plateau's short-horned lizard, Bell credited the source of his information:

This beautiful and highly interesting species was found by Mr.
David Douglas in the course of his late indefatigable and productive
researches in the western parts of North America, to whom I am
also indebted for the following account of its habits.

Bell quoted Douglas extensively on the behavior of this scaly reptile, which the collector had found in the burrows of jackrabbits and ground squirrels, always within a mile and a half of water. The naturalist had dissected several of the lizards, easily captured during the cool days of spring and fall. "The traveler is constantly annoyed by them during the night, seeking shelter from the cold under his blanket, and is frequently under the necessity of removing these little intruders on his rest," Douglas noted. Upon examining the contents of the stomachs of these little intruders, he had concluded that they were omnivores who fed both on beetles and on the leaves of desert shrubs. He had also observed perfectly formed young, half an inch long and equally nimble as the adults, in early spring.

The article was enhanced by Bell's exquisite drawings of the strange creature, with every scale minutely detailed, its body sinuously curved in a semblance of life. Even though some of the behavioral traits reported by Douglas are questionable—modern herpetologists have found horned lizards far from water on the Columbia Plateau, and never remember shaking any out of their sleeping bags—the collector certainly deserved the honor of the species name *douglasii* bestowed by Thomas Bell.

At this same meeting of the Linnean Society, Douglas was officially elected a Fellow. Although new fellows were usually charged an induction fee of twenty-five pounds and an annual subscription, the council had "ordered that all Fees payable on the Admission of Fellows, and all Fees payable in future, be remitted in the case of Mr. David Douglas, in consequence of the great services he has rendered to Natural History." Other prominent scientists continued to salute his contributions. Ornithologist Nicholas Vigors, attaching Douglas's name to a species of grouse brought back from the Northwest, added: "Few individuals have done more to elucidate the Natural History of any distant country than Mr. Douglas."

Douglas's interests were not entirely confined to the realm of natural history, and he became fascinated with the research of Edward Sabine (Joseph's brother) regarding the Earth's magnetic poles. Ten years older than Douglas, Edward had served as an officer in the Royal Navy during the Napoleonic Wars, and more recently had accompanied John Ross's expedition to the Arctic. Captain Sabine was a serious amateur scientist with an obsession for accurate navigation, and during the summer of 1828, he invited Douglas to assist him in experiments with the dip needle, a pivoting compass designed to measure the radial, longitudinal, and latitudinal components of the Earth's magnetic field. The captain found his helper "so apt and so grateful a pupil, that a cordial friendship was established, which continued to the last."

In early September 1828, Douglas boarded a coach to Glasgow to visit Hooker. Gifts for his mentor included samples of wood from Northwest trees and a curious little book called *Molecules* by their friend Robert Brown describing the mysterious erratic disturbances of plant pollen that would come to be called Brownian motion. Douglas had also obtained the Horticultural Society's permission to share some of his extra dried specimens.

"He has brought a complete sample set of all his things for me," Hooker wrote, "with the exception of 2 or 3 species which have been rather mysteriously lost. They are indeed excellent things." Hooker was then in the process of compiling his *Flora Boreali-Americana,* an ambitious manual on the plants of British North America. He naturally hoped to include all of Douglas's significant finds, and with the first volume nearing completion, he relished the opportunity to quiz the collector in person. He urged Douglas to complete his own descriptions of his collections, but had to admit that his protégé was "never so unhappy as when he has a pen in his hand."

Escaping from their pens, the pair visited some of their old haunts in the Highlands accompanied by Hooker's son Joseph Dalton, who was ten years old that year and quite excited about fishing. Writing to the professor a few weeks later, Douglas asked him to tell Joseph "that a fisherman was on the coach that brought me to town, had three trout tied in an old cloth, but they were not by any means so fine as our Ben Lomond ones."

During this trip, Douglas also traveled north to his child-hood home in Scone for a visit with his mother and two of his sisters. According to the recollections of a neighbor, his eyes were bothering him considerably at the time, and his manner remained somewhat subdued. This family friend later summarized the village's attitude toward their local lad:

> *Faults Mr. Douglas may have had—for humanity in its best form is not free from them, but his friends and family saw them not. Well might they love him sincerely—for in every respect, in his palmist days, in his attentions and kindnesses as a son and a brother, he might be equaled, but he could not be exceeded.*

At some point during Douglas's stay in Glasgow, Professor Hooker arranged for him to sit for a portrait by artist Sir Daniel Macnee. In the completed painting, Douglas's head rises above two sharply cut lapels, and a white linen collar winds up to his dimpled chin. He sports an alert, determined look, with wavy thinning hair and deepset, heavily shadowed eyes. His healthy pink cheeks and gray brows are creased with the markings of an outdoor life, and Macnee cast a noble light across the right side of his subject's face.

Around this same time, Douglas's niece also convinced him to pose for a pencil sketch. Although he is seated in much the same pose, she drew her uncle in less formal dress and with none of the noble shadows and clouds. In his niece's sketch, Douglas's smooth chin droops a bit; the contours of his face are less pronounced. He looks like a small, balding man who would never be able to enjoy sitting still.

Grouse and Gooseberries

Upon his return to Chiswick that November, Douglas contracted a severe cold and sore throat that hung on for almost two weeks. Despite his illness, he replied to a query from R. W. Hay of the Colonial Office regarding the western portion of the international border between Great Britain and the United States. The Treaty of Ghent's ten-year agreement of joint occupancy in the Pacific Northwest was drawing to a close, and the British Colonial Office was gathering information to help construct their case for the placement of the boundary line. Hay asked Douglas, who had certainly seen more of the Columbia drainage than anyone in England, for his assessment. His reply described the river from its mouth to Boat Encampment, extolling the vast area's valuable resources and mild climate. In regard to the boundary, he agreed with the Hudson's Bay Company recommendation to mark Lewis and Clark's route across the Rockies at 46° of latitude, then trace the Clearwater, Snake, and Columbia rivers to the sea.

By mid-November Douglas was sufficiently recovered to attend the monthly meeting of the Linneans, where he met Francis Boott, a Boston physician and botanist who had recently moved to London. "Having no one to introduce me," he explained to Hooker, "I did it myself, and I say that it was no difficult task." While dining with Boott's family the following week, Douglas had a look at the Bostonian's herbarium. "Who can dry specimens like him?" he exclaimed. "Once did I think I could make good ones, but a sight of his put me speechless."

At the next meeting of the Linnean Society in December, Douglas read a paper on North American grouse and quail. He had reported to Hooker that he was "no longer lazy, working hard when able," and he had obviously spent considerable time compiling information on these birds from his field journals. His account began with a lengthy treatment of sage grouse, referred to as the "Cock of the Plains" in Lewis and Clark's descriptions. Douglas covered every facet of their life history, from their varied diet to their Cayuse name, *py'amis*; from the irregular chocolate splotches that graced the thick ends of their eggs to their courtship dances:

> The wings of the male bird are lowered, buzzing on the ground, the tail spread like a fan, somewhat erect; the bare yellow oesophagus inflated to a prodigious size, fully half as large as his body, and from its soft membranous substance being well contrasted with the scale-like feathers below it on the breast, and the flexile silky feathers on the neck, which on these occasions stand erect. In this grotesque form he displays in the presence of his intended mate a variety of pleasing attitudes.

He continued in the same vein with treatments of ruffed, spruce, and dusky blue grouse. Turning his attention to quail, he compared the topknots of California and mountain quail, remarked briefly on ptarmigan and prairie chickens, then overreached himself as he tried to separate half a dozen suspected species observed on his journey from the Rocky Mountains to Hudson Bay. He concluded with a promise of more novelties to come: "I look forward at no distant period to again resuming my labours on the western parts of the same continent, the result of which, in due season, it will afford me the greatest pleasure to submit to the Society."

Douglas was not the only person wishing that he would return to the grouse's dancing grounds. "His best friends could not but wish, as he himself did, that he were again occupied in the honourable task of exploring North-west America," Hooker later wrote, comparing Douglas's happiness while "surveying the wonders of nature in its grandest scale" with the restlessness and dissatisfaction that plagued him after his return to England. The professor added that the collector's initial pleasure at being back in his native land had been superceded by "many, and to his sensitive mind, grievous disappointments."

Douglas had complained in late fall that "the London climate kills me," and over the winter his mood took a nasty turn as well. His former feelings of friendship toward John Lindley, for example, were replaced by resentment over the usage of his Northwest collections. After a November dinner at which Lindley delivered an impromptu speech, Douglas caustically commented: "The beginning was bad, the end was bad and the middle worthy of the beginning and the end . . . the manner of delivery shockingly ill."

As time passed with no prospect of another expedition, "his temper became more sensitive than ever," Hooker observed. His quarrelsome tendencies resurfaced. William Booth, who saw him daily, tried to justify his friend's behavior: "Soured in his temper . . . and finding that, after all his exertions, he was not even so well paid as the Society's porter on Regent Street, he naturally grew restless and dissatisfied." The secretary of the Zoological Society touched on this sore subject when he made the innocent mistake of adding an "Esq." to Douglas's name on a membership list. Perhaps stung by some recent class slight or the pinch of monetary woes, the collector fired off a protest, insisting that the title "Esquire" did not fit him in any way. "As it is doubtful if I even can afford a residence I have further to request that you give me no address," the disgruntled Douglas concluded. If he couldn't be somebody, he would insist on being a nobody.

The arrival of spring 1829 brought the welcome respite of a visit from William Hooker, who traveled to London in late April to obtain the Horticultural Society's permission to publish data from Douglas's journeys in his forthcoming *Flora*. (As sponsors of his expedition, the Society controlled the rights to all his material). Joseph Sabine, who frequently shared specimens with the professor, gladly granted Hooker's request. Douglas, who had previously proposed giving all his information to his mentor, would certainly have acquiesced.

If Hooker was still in London for the Horticultural Society's April 21 meeting, he would have been able to observe his protégé's presentation of "An Account of some new, and little known Species of the Genus *Ribes*." As had become his habit, the author began with the star of the bunch: *Ribes sanguineum*, the showy red-flowering currant of the Pacific Northwest. He had sent its seeds to the Society in the fall of 1826, and now he proudly announced that "they blossomed in great profusion, though scarcely two years old." In this case Douglas's pride was well-founded, for the red-flowering currant soon proved immensely popular with English gardeners. John Lindley maintained

> of such importance do we consider it to the embellishment of our gardens, that if the expense incurred by the Horticultural Society in Mr. Douglas's voyage had been attended with no other result than the introduction of this species, there would have been no ground for dissatisfaction.

The Instruments of Return

With his spirits buoyed by Hooker's visit and his successful *Ribes* presentation, Douglas reported that he was again focused on his neglected narrative: "I am doing my best to get finished." He was also devoting energy to ordering his specimens, stating with satisfaction, "I have got my Herbarium arranged and placed in the Council room . . . very neatly."

That June, his herbarium and living collections were exhibited at the Horticultural Society's third annual fete. This festive fundraiser, hailed as "one of the principal attractions of the fashionable season," featured floral arrangements and garden tours in addition to elegant refreshments, musical performances, and dancing. By mid-afternoon on the appointed day, more than three thousand "persons of the first respectability" were promenading past tents and marquees, admiring, among other displays, the first flowering of Douglas's beautiful pink clarkia. Suddenly the festivities were interrupted by one of the heaviest rainstorms of the season, with the *Morning Post* reporting that the ensuing "shrieks were dreadful, and the loss of shoes particularly annoying!"

Douglas, one of the few attendees who would not have been annoyed by muddy shoes, soon experienced a favorable change in the weather regarding his own affairs: in mid-July, the Horticultural Society's council voted to send him on another expedition to the Northwest. The Hudson's Bay Company "made a most liberal offer of assistance," promising a berth on their fall supply ship to the Columbia. He would receive £120 per year, a raise of £20 over his previous salary. Douglas made arrangements to have one-third of his earnings forwarded to his brother John to help with the care of their mother and sisters.

Over the course of the summer, a rejuvenated Douglas communicated frequently with Hooker, describing himself as busy and productive, and his temperament much improved: "I have had only or two *very slight outbreakings,* as Mr. Sabine calls my fits, since I saw you." His enthusiasm was almost palpable on the page, and his self-effacing humor resurfaced. His eyesight, however, remained problematic. In early August, he reported that he could no longer read small print, and asked Hooker to send him a Bible with "good bold legible type." He had been unable to find one in London, but recalled a Scottish edition that would suit his needs.

Upon learning of Douglas's impending expedition, R. W. Hay of the Colonial Office offered to cover the principal part of the collector's expenses abroad in exchange for geographical observations relating to the proposed boundary. He also promised to pay Douglas personally for charts and other information upon his return home. This deferred payment made sense to Douglas, who joked to Hooker that "if I had a good salary, I might fold my hands and become lazy, therefore I can feel no objection to being paid according to my labour."

The Colonial Office assignment required a cram course in surveying and practical astronomy, so in late August Douglas journeyed to Greenwich to place himself under the tutelage of Edward Sabine. "I heard him frequently express his regret that his limited education prevented his being able to render those services to the geographical and physical sciences," Sabine wrote. "I told him, that if his energy of purpose was equal to the strength of the wishes he expressed, I was of the opinion that it was possible for him to acquire the requisite knowledge." The captain later boasted that his pupil studied eighteen hours a day, and "conquering every difficulty, . . . acquired a competent knowledge of the principals of science." Douglas took his surveying responsibilities very seriously, not only for the services he might render the Colonial Office but also for the advancement of scientific knowledge in general. This would not, he insisted, be "the journey of a commonplace tourist." He planned to use a hygrometer to measure the relative humidity at various altitudes and latitudes "which will effect much in illustrating the Geography of Plants."

Through his naval experience, Edward Sabine was acquainted with a variety of scientifically minded sea captains, and he introduced his student to a visiting Russian skipper named Fyodor Lutke. Conversation with Captain Lutke fueled Douglas's long-simmering dream of an expedition into Russian territory.

> When I have completed my expedition on the Continent of America, I may cross to the opposite shore, and return in a southerly line, near the Russian frontier with China. What a glorious prospect! Thus not only the plants, but a series of observations may be produced, the work of the same individual on both Continents, with the same instruments, under similar circumstances, and in corresponding latitudes! I hope I do not indulge my hopes too far.

Lutke assured the collector that his hopes were attainable, and backed up his encouragement with letters of introduction to the governor of the Aleutian Islands and to Siberian officials on the far side of the Pacific.

By mid-September, after completing his preliminary instruction in the practical use of a compass, chronometer, thermometer, barometer, hydrometer, hygrometer, dip needle, and sextant, Douglas was happily engaged in final preparations for his journey. "I ought to think myself a very lucky fellow," he wrote, "for indeed every person seems to take more interest than another in assisting me." He had applied to the Colonial Office for £60 so he could obtain surveying books and charts, and Undersecretary Hay sent along an extra £20, along with a set of instruments that, though used, were in perfectly good order. Edward Sabine took Douglas to John Dollond and Sons, the premier instrument makers in London, to assemble a more complete kit worth the eye-popping sum of £280. The Zoological Society donated a double-barreled shotgun worth £18; the editor of their *Journal* predicted that "we may now anticipate some valuable information respecting his zoological researches, information which will raise him as highly in reputation as a scientifick naturalist, as he at present stands as an enterprising traveler."

While preparing for his foray into chart-making, Douglas reviewed a proof of the large map that would be tipped into the first volume of Hooker's *Flora*. Douglas's western travels were traced on the chart alongside those of recent Arctic expeditions. "The route of Franklin, Richardson, and Drummond is marked in RED, Parry's in BLUE, and mine in YELLOW," Douglas wrote to Hooker, noting that although the map was very fine, he did not approve of the engraver's choice of colors. "I must have the latter tint changed to green, for yellow is a most sickly hue for a *culler of weeds*." By the time the volume

was published, Douglas's circuitous pathway had been trans-
formed into a more appropriate green.

That pleasingly lush line followed the collector's journey
across Athabasca Pass, between the newly named Mounts
Brown and Hooker. Douglas, who had no practical surveying
knowledge when he topped the Continental Divide, marked
their elevations as 16,000 and 17,000 feet—too tall by more
than a third. In this he was following the estimate of David
Thompson, who had pioneered the pass in 1811 but did not
carry the equipment necessary for an accurate measurement.
The mistake was soon copied, and lingered on maps for the
next half century.

During that fall of 1829, talk around the Hudson's Bay
Company offices focused on the wreck of the *William and Ann*,
which had gone down the previous spring while attempting to
cross the Columbia bar. Even more disturbing was the rumor
that some of the forty-six survivors had been dispatched by
local Chinook people. "This may be true," Douglas wrote upon
hearing the report, "though I confess I entertain some doubts,
for I have lived among those people unmolested for weeks and
months." (The collector's intuition proved correct, for John
McLoughlin visited the scene and exonerated the Chinooks of
any wrongdoing beyond appropriating most of the cargo that
washed ashore.) In the voice of a man who had succeeded in
making himself at home on the Columbia, Douglas asserted,
"I shall yet venture among these tribes once again. . . . I doubt
not if I can do as much as most people, and perhaps more than
some who make a parade of it."

Douglas's letters to Hooker were often sprinkled with sharp
remarks, and his reference to men who paraded their influence
hints at some dissension during his visit to the Northwest.
Dr. John Richardson, who had an ear for gossip, had heard

rumors among the Hudson's Bay Company traders during his journey across Canada in 1827. "Douglas made himself some enemies in the Columbia," he confided to Hooker, "but he has not hesitated to go back to that quarter again & now that he knows the characters of the men with whom he has to deal will do better."

While awaiting the summons to sail on the Hudson's Bay Company ship *Eagle* in late October, Douglas received a copy of the first volume of Hooker's *Flora*. He quickly penned a congratulatory word to the author: "I cannot tell you how pleased I am to have seen the first part before sailing, and that I am enabled to take it with me to America. . . . The type is also good, and the notices and habitats full—a point of great importance."

On his title page, Hooker acknowledged that the fruits of the book were

> *compiled principally from the plants collected by Dr. Richardson and Mr. Drummond on their late northern expeditions.*
>
> *To which are added*
>
> *(by permission of the Horticultural Society of London,)*
>
> *Those of Mr. Douglas, from North-West America, and of other Naturalists.*

When he had time to peruse the volume, Douglas would have found his own name celebrated in the very first entry: "This beautiful species of *Clematis* (*C. douglasii*) is quite unlike any hitherto described; and I am anxious it should bear the name of its zealous and meritorious discoverer." Beyond that clematis, Douglas could have reveled in numerous snippets and ethnographic comments extracted from his journals, letters, and conversations with Hooker. Perhaps equally gratifying to Douglas would have been Hooker's acknowledgment of his technical proficiency, retaining many of the collector's species designations. Douglas, for his part, was looking to the future.

> I hope, ere the whole of the Flora is printed, to be able to supply you with many and striking novelties. I am sensible of the great advantage I derive from my former experience of traveling in the country . . . and certainly if I find the Indian tribes as quiet as when I left them, much good may be effected.

Regarding the publication of his own book, Douglas remarked: "I have applied to Mr. Sabine for a copy of the sketch of my journey, but have not yet obtained this. I shall or he must publish it soon." Whether Douglas meant that he intended to keep working on his manuscript, or that Sabine was holding it back for some reason, is unclear, but he obviously had not given up on publishing an account of his Northwest travels. He had much of value to offer, and his second expedition would surely add to that store. "I shall write every day, and write everything," he promised Hooker, "so that my driveling will come home, if not myself."

X.
Breathing New Climates
Fall 1829–Fall 1832

His Enlivening Society

D ouglas's journals from his second journey to the North-west were lost, and letters to friends and associates contain his only descriptions of events after he set sail from London on the last day of October 1829. His surveying data did survive, establishing his locations for certain days, and boxes of plant specimens that he shipped to England provide hints to the habitats he visited. But there are no detailed records of the flora and fauna he collected, and no day-to-day accounts of his adventures, so breathtakingly related during the course of his first journey.

"During the whole passage my days were only moments," Douglas wrote to Edward Sabine from the *Eagle*. As the ship plied the South Atlantic, he continued his studies in practical astronomy, familiarizing himself with each item in his kit. "How beneficial it is for a person like me to be at sea some months previous to engaging on a long journey. I have had time to think over and settle my plans; and I never suffered an opportunity to pass without endeavoring to perfect myself in the use of some instrument." He diligently recorded his observations on lithographed forms that Sabine had provided, taking care to position himself as far as possible from iron lest the metal affect his compass or dip needle. While he lamented the *Eagle's* failure to call on any ports along the eastern shore of South America, where he had anticipated making new collections, her straight run to the Pacific allowed him to keep his focus on measurements of position and magnetic variation.

On her way across the Pacific, the *Eagle* did make a brief layover in the Sandwich Islands (as the Hawaiian Islands were then known). She dropped anchor at Oahu during the rainy season, when the only plants that Douglas could preserve were mosses and ferns, so he was unable to take a full measure of the new landscape. He did, however, perform one "service to Botany, in scaling the lofty and rugged peaks of Mouna Parrii" (possibly Mount Kaala, the island's highest point). He thought the tropical verdure held great promise, and assured Hooker that if he were able to return at a more favorable time of year, he could collect enough material to warrant a full volume on Hawaiian flora.

Equally interesting were the people Douglas met on Oahu. Several members of Hawaii's royal court had visited England in 1824, and now welcomed a British traveler to their country. These included the young Royal Governess Liliha Boki, who entertained him with great style. Before he departed, Douglas shipped a collection of fine books, printed on the island and "splendidly bound in tortoise-shell," as a gift for Professor Hooker's library.

Back aboard the *Eagle*, Douglas again turned his attention to celestial observations, tracking the daily increase in latitude toward 46° North. As the ship swept across the Columbia's bar in early June 1830, he beheld an alarming sight—the wreck of the Hudson's Bay Company vessel *Isabella*, on which he had originally been scheduled to sail. Upon reaching the Columbia a month earlier, she had run aground near the same sandbar where the *William and Ann* had gone down two years before. Although the captain and crew had escaped safely, Douglas shuddered at the thought that if he had been on board the *Isabella*, "I should have lost my all; and think what a plight would mine have been, on 'Cape Disappointment,' deprived of every thing!"

In a letter to Hooker, Douglas remarked matter-of-factly that when he left England, "some of my fashionable London friends . . . were glad to get rid of me." Far different sentiments were expressed by the residents at Fort Vancouver upon his arrival there—by this time he was acquainted with nearly every agent serving in the Columbia District, and had become a favorite with several of them. "We were glad to see again amongst us an old friend, with his noble countenance, and agreeable hearty manner unchanged," wrote George Barnston.

Douglas would have immediately noticed that great alterations had taken place at Fort Vancouver during his three-year absence. The entire post was being moved from the bluff down onto the flats for better access to the river. New features included a boatyard and a sawmill four miles upstream. John McLoughlin had greatly expanded the farm, and the crops for that year included wheat, barley, Indian corn, oats, and three different kinds of dried peas.

There had been political changes as well. The American ship *Owyhee* from Boston was moored just downstream from the post, and the competition for trade had created tension among the tribes along the lower river. The Bay Company's relations with several of the coastal tribes had deteriorated alarmingly. In January 1828, five men traveling between the lower Fraser and the Columbia had been killed by members of the Clallam tribe from Puget Sound during an apparent robbery attempt; among the victims was Alexander McKenzie, Douglas's traveling companion in 1825 and 1826. A retaliatory party had killed twenty-two Clallams and burned their entire village—a brutal action that greatly displeased the governors in London.

Douglas would have heard versions of this and other happenings as he arranged his gear. George Barnston, "pleased to find that his stature as a disciple of science had greatly increased," marveled at the quality of his friend's instruments, especially a fine asbestos thread for suspending magnetic bars from a tripod, supplied by a famous French astronomer. Douglas immediately busied himself with testing his equipment, taking observations to calculate the coordinates of Fort Vancouver, but Barnston made sure that "during his leisure hours we had the enjoyment of his enlivening society."

Near the end of June, when the interior traders set off for their various posts with fresh supplies, Douglas hopped aboard one of the canoes. Barnston, who had been posted to Fort Walla Walla that season, declared the collector to be an enlivening companion on the water as well as on land: "He would frequently spring up abruptly in an excited manner, and with extended arms keep his finger pointed at a particular spot on the beach or the shelving and precipitous rocks, where some new or desirable plant had attracted him." When the voyageurs indulged their passenger and maneuvered the canoe close to shore, "we would then be amused at the agility of his leap to the land, and the scramble like that of a cat upon the rocks . . . happy if he achieved this without slipping, and falling into the deep water alongside the boat."

Douglas's vision apparently had improved, for he displayed a remarkable quickness of sight "in the discovery of any small object or plant on the ground over which we passed." The exuberant collector, aided by summer's clear skies and Barnston's experienced hands, also amassed geographic data at every opportunity, taking multiple readings and calculating the coordinates and angle of declination at key landmarks such as the Dalles and the mouth of the Snake River. "My instruments are all excellent," he wrote to Hooker, "and have already enabled me to make a multitude of important observations, which will go some way toward perfecting the Physical Geography of this part of the country, as well as illustrating its magnetic phenomena."

When Barnston disembarked at Fort Walla Walla to assume his new command, Douglas paused there as well. Intent on a fourth expedition into the Blue Mountains, he planned to compare the altitudes of the peaks with plant ranges along their slopes. Barnston outfitted him with an interpreter, a sturdy boy, and five horses to pack the bulky instruments. Although most of the spring flowers were already withered by summer's heat, Douglas did find some prized Brown's peonies whose seeds were perfectly ripe. As on his previous journeys, he hoped to continue south from the Blues to explore the Umatilla and Grande Ronde country, but, as had happened each time before, his guide's fear of Shoshones scuttled his plans.

This setback failed to dampen the collector's spirits, and upon his return to Fort Walla Walla, he enlisted Barnston's help in collecting lizards. They hired some of the Indian boys who lived near the fort to crawl about and search for the small holes that marked the entries to the "reptile's apartments." Armed with horsehair lassos attached to long noosing sticks, the boys would throw themselves flat on the hot sand beside a promising hole until their quarry showed his head. Then "they would quickly suspend him with one jerk, and bring him like a culprit to our sides: a slight reward would put them in ecstacies, and they would again scamper off for renewed captures." The boys' captives included not only the short-horned species that Douglas had given to Thomas Bell, but also a "beautiful long-tailed little lizard" with iridescent blue flanks—either the male sagebrush or the western fence lizard.

After Douglas departed with a canoe headed downstream on July 23, Barnston "felt his absence as a sad blank." The feeling was mutual, especially when Douglas paused for more astronomical observations.

> I wish you had been with me to have lent me a hand, for I had some trouble boiling the mercury in the tube. Fortunately I can find only .004 of an inch of index error, from the comparison I made with it and my others at Greenwich. I could have done no more, had I been in Dollond's shop.

When the group pitched camp near the Cascade Rapids on July 25, Douglas set off in search of Chumtalia, hoping to convince his former guide to accompany him on another trip into the mountains. Upon reaching the village, he found a "hyass patchatch, or great feast," in progress, with visitors from distant parts who had gathered to help Chumtalia celebrate "the never-to-be-forgotten occasion of the perforating the septum of his young daughter's nose, and piercing her ears." Realizing that "it would have been very ungallant to the young lady, as well as ungracious toward the father, to have pressed him to go with me," Douglas postponed his mountaineering plans.

Continuing downstream to Fort Vancouver, he watched the kale harvest and greeted Alexander McLeod, his compatriot from the Umpqua country. Since their last meeting, McLeod had been involved in a series of controversies, including the punitive 1828 mission against the Clallam tribe to avenge Alexander McKenzie's murder. During subsequent journeys to the southward, McLeod had displeased John McLoughlin by disregarding orders and mismanaging supplies. Despite his strained relations with the company, McLeod retained charge of the Umpqua expedition, and when he departed the post in early August, Douglas joined him for the first part of the trip through the Willamette Valley.

John McLoughlin, realizing that Douglas needed assistance to carry out all the measurements he hoped to take, supplied him with a "personal servant"—a cowherd and trapper named William Johnson, whom the collector had met on the Umpqua expedition in 1826. Long before that, Johnson had served as a common sailor in both the British and American navies, and delighted in recounting his adventures from the War of 1812.

With this "man o'war's man" as his assistant, Douglas set off for the familiar landscape of the Willamette Valley, trailed by "my faithful little Scotch terrier, the companion of all my journies." Whether this terrier, named Billy, was the same dog that had warmed Douglas's feet on his previous journey or whether he had brought him over on the *Eagle* is unclear, but the pet was seldom far from his master's side.

Making surveys as he traveled, Douglas set up his equipment to triangulate an accurate height for Mount Jefferson, but "the day however was not so good as I could have wished, the snow capped summit was obscurely defined by reason of flaky and stray milky clouds, that adhered to it with great obstinacy." His calculation of Jefferson's altitude was too high by about 10 percent, well above the margin of error that Edward Sabine thought he should be achieving.

During his three-week sojourn down the Willamette with the loquacious Johnson, Douglas would have learned about much that had transpired in the "southern country" over the last two years. Johnson and McLeod had survived a hostile encounter with the Umpqua tribe, and both had returned to that river to recover the property of American trapper Jedediah Smith, who lost most of his crew and all of his pelts after a dispute with an Umpqua leader in 1828.

But the Kalapuyas along the Willamette remained friendly, and Douglas and Johnson traveled with ease. In addition to his geographic data, the naturalist steadily accumulated collections of flora and fauna, and just as steadily lost them. While he was fording the Santiam River, a packet of his recent zoological finds washed away, "a dreadful loss, as it contained good things." Many of his Umpqua specimens had been destroyed in this same river in November 1826. "It is curious," he mused. "A kelpie or elf is the charm of that stream, so unfortunate to me."

Only a few days after Douglas returned to Fort Vancouver, Chumtalia arrived in his large canoe, and the collector accompanied him upstream. Together the pair must have retraced their 1825 trek up the shoulder of Mount Hood, because Douglas returned to the post with samples of noble fir, the tree Chumtalia had shown him there previously. But this success was soon spoiled by the news that "poor Chumtalia is since dead. He was blown up by his powder horn which was on his person, and falling on his side, his knife entered above the fifth rib so that he died."

Within a few weeks, news of death became commonplace, for a terrible epidemic had appeared on the lower Columbia. Known as "fever and ague" or "intermittent fever," it swept along the river between Fort Vancouver and the coast. By the end of September, more than half the people at the post were afflicted, including the new physician. "To attempt to describe our situation during this dreadful visitation is impossible," John McLoughlin later wrote of the weeks during which he and two clerks tended the sick from daybreak until midnight. According to George Barnston, "Douglas was ill like others, but being something of a leech, had an early recovery, and recruited perfectly by following up his wonted healthful perambulations." Barnston must have been alluding to the definition of *leech* as a physician, in reference to Douglas's courses under Hooker. His remedies apparently worked, for the "leech" described himself as hovering for ten days between hope and fear, without ever being completely incapacitated.

> *I am one of the very few persons among the Hudson Bay Company's people who have stood it, and sometimes I think, even I have got a great shake, and can hardly consider myself out of danger, as the weather is yet very hot.*

While most of the Canadian workers and mixed-blood families at Fort Vancouver recovered, the disease wreaked havoc among the Hawaiians and the native population. Large numbers of Indians from surrounding villages came to camp near the fort, "giving as a reason that if they died they knew we would bury them." Yet the traders were soon unable to render even that small service, for the sheer volume of deaths proved overwhelming. "Villages, which had afforded from one to two hundred effective warriors are totally gone," Douglas wrote to Hooker in early October; "not a soul remains! The houses are empty, and flocks of famished dogs are howling about, while the dead bodies lie strewed in every direction on the sands of the river."

Modern epidemiologists consider intermittent fever to be a strain of malaria brought in by ships that had picked up infected mosquitoes during their passage through the tropics, but Hudson's Bay Company officials at the time remained mystified as to its origin. Barnston wondered whether the cultivation of a hundred new acres of land around Fort Vancouver might have released the plague. Some of the tribes were certain that the American captain of the *Owyhee* had poured the infection into the river from a bottle he kept on board; others attributed the scourge to buoys that the sailors had placed to mark the channel.

By early October, Douglas was sufficiently recovered from his bout of fever to pack three chests to send to England on the *Eagle*. One packet contained seeds of six fine conifers, including grand and noble firs. Propagated at Chiswick, the noble firs quickly became a favorite in manor house landscapes, and within a few years, healthy sprouts were selling for fifteen to twenty pounds apiece.

With the *Eagle*'s sails unfurled and the captain standing by for a last bit of mail, Douglas packaged a sampling of mosses for William Hooker and dashed off a letter outlining his future plans. "My desire is to prosecute my journey in North California," he wrote. "If I can venture thither in safety by land, I will do so; if not, I shall go by sea to Monterey."

Once again, his wishes coincided with company business—the vessel *Dryad* was due to depart soon carrying a load of sawn timber and salted salmon for sale in California. Since Douglas planned to spend a considerable time collecting there, he assembled a stock of personal items to see him through. From the Fort Vancouver storeroom, he charged to his account extra shoes, shirts, cloth, cod line, and candles, along with ten pounds of Hyson tea, sixty pounds of loaf sugar, nine gallons of Madeira wine, four silk handkerchiefs, a Jew's harp, and 150 Spanish dollars. Funding, it appears, was not as much of an issue as it had been on his earlier visit.

In late November 1830, the *Dryad* weighed anchor at the fort and sailed down the Columbia. Douglas and William Johnson set up their instruments, taking observations on Fort George and Cape Disappointment as the ship passed out of the river and into the Pacific. The collector didn't know how long he would be gone, but he carried a voucher that would provide him with free passage when he was ready to return:

To the Commander of any of
The Honble HB Coy Vessels
Bound to the Columbia River

Sir
You will receive on Board your Vessel Mr
David Douglass with his Servant and Baggage
Yours truly
J McLoughlin

California

On December 22 the *Dryad* dropped anchor in Monterey Bay in the newly established republic of Mexico. Douglas introduced himself to William Hartnell, the unofficial British consul of Monterey, with a letter from a British sea captain he had met in London.

Mr Dear Sir

Permit me to introduce to you Mr. Douglas, an enterprising friend of mine who has already made one Journey across the Continent of America, and is about to undertake another. . . . Any attention you can pay him or assistance you can render him in his arduous undertaking will greatly oblige yours very truly.

F. W. Beechey

Hartnell obliged the captain by ushering Douglas to his comfortable villa perched on a hill overlooking the stucco buildings of the village and the abundant avian and marine mammal life that plied the long arc of the bay beyond. Hartnell was known as a generous host and an accomplished linguist who moved comfortably within the Spanish world; he had married the daughter of a presidio commander, and the couple often housed foreign guests for long periods of time. Douglas, who carried in his pocket a small Spanish dictionary that Hooker had given him six years earlier, would have found ample opportunities to improve his vocabulary in the bilingual Hartnell household.

Douglas had the singular fortune to disembark at Monterey on the winter solstice, just at the beginning of the rainy season. That allowed him to synchronize his rounds with the rapid succession of flower, fruit, and seed that takes place in a two-season climate. He had a further stroke of luck when the Mexican officials delayed issuing a travel visa, confining him to the fertile ground around Monterey Bay for three months.

While grousing about the glacial pace of officialdom, he took stock of the luxurious plant growth. Almost as soon as he touched land in Monterey, he found a fuchsia-flowered gooseberry, cousin to the beautiful red-flowering currant. Close by, he stumbled upon a charming ground cover, "the harbinger of Californian spring, which forms as it were a carpet of azure," known today as baby blue-eyes. The collector must have received conditional approval to travel a certain distance, for he visited the mission at San Juan Bautista in early February 1831, where he found the large-coned, sprawling oddity from mountain foothills now called digger pine. Apparently the passage of time had assuaged his hurt feelings toward Joseph Sabine, for while staying at the mission, he wrote a short notice about the tree, naming it *Pinus sabiniana*, "well meant on my part and conferred on one who truly deserved it." In time Sabine himself would read the description before the Linnean Society. "I am pleased to know my old friend, Mr. Sabine, thought his Pine worthy of him," Douglas later wrote to Hooker.

A few weeks later, Douglas traveled north to Santa Clara. Along the way he found five new mariposa lilies plus a blazing-star he named for John Lindley. He measured gigantic redwoods with the same foreboding awe that he had sensed in the coastal rain forests of the Northwest: "The great beauty of the California vegetation is a species of *Taxodium* [*Sequoia*] which gives the mountains a most peculiar, I was going to say awesome appearance—something that plainly tells us we are not in Europe."

Douglas made many friends within the small foreign community in Monterey and became something of a local celebrity. Alfred Robinson, an American businessman who traveled up and down the California coast, met the collector in 1830. "I found a new resident at Monterey—David Douglas, Esq., a naturalist from Scotland . . . I was told he would frequently go off, attended only by his little dog, and with rifle in hand search the wildest thicket in hopes of meeting a bear; yet the sight of a bullock grazing in an open field was to him more dreadful than all the terrors of the forest."

According to California pioneer William Heath Davis, Douglas had a reputation as a medical man as well as a naturalist. During a visit to Monterey when he was nine years old, Davis fell through an open hatch cover on a ship and broke his arm. Someone summoned help, and "Dr. Douglas set my arm carefully, and treated me very kindly . . . he was a good, good man." The set bone healed cleanly and never troubled Davis for the rest of his long life. Douglas became known locally by the title, "Doctor," which eventually followed him to Hawaii.

When his traveling papers were finally stamped in April, they included a provision forbidding him from sketching any military fortifications—not a particularly surprising caution, given that he was carrying an extensive assortment of surveying equipment. With his visa in hand, Douglas moved farther afield, sweeping up everything around him, then employing his full array of instruments to ground his collections. "Their localities are determined," he declared, "altitudes measured, the climate they breathe analyzed."

As Douglas and William Johnson moved south along El Camino Real with their tools in tow, Franciscan missions served as the same sort of safe havens that fur trade posts had on the Columbia, though much more conveniently situated about a day's ride apart. "I lived almost exclusively with the fathers who without an exception, afforded me the most essential assistance, hospitality to excess, with a thousand little courtesies which we feel and cannot express," Douglas wrote. The padres did not evangelize, but "gave me always a good bed, and plenty to eat and drink of the best of the land." He found them to be well educated—versed in Latin and quite understanding of his eccentric occupations: "They know and love the sciences too well to think it curious to see one go so far in quest of grass."

The pampered traveler entered the coordinates of San Luis Obispo in his survey notebook in early May and those of Santa Barbara by the middle of the month. It wasn't long, however, before the dry season curtailed his collecting. "The intense heat set in around June, when every bit of herbage was dried to a cinder. . . . In this fine district, how I lament the want of such majestic rivers as the Columbia!" The bright sunlight bothered Douglas's weak eyes so that reading his instruments became a strain. He retreated to Monterey by the end of the month, but soon headed north to explore around San Francisco Bay, reaching San Rafael on July 27 and two days later visiting the mission at the modern city of Sonoma. His party apparently pushed west to the sea and the Russian outpost at Fort Ross, whose position he calculated as 38° 45' N. One of Douglas's hopes was to revisit the Umpqua country, so he could connect his travels north from Monterey with his 1826 excursion south from Fort Vancouver. He later told Hooker that he came within sixty-five miles of achieving that goal, but his survey book shows nothing beyond his entry near Fort Ross, more than three hundred miles south of the Umpqua—a considerable error in estimation.

From his first year in California, Douglas sent to London about the same number of plants—five hundred—as he had during his initial visit to the Northwest. Although he thought that number "vexatiously small" compared to the riches of the region, he was ready to move on. He expected the *Dryad* to arrive shortly and deliver him back to the Columbia, which could provide the jumping-off point for his next great adventure. He had recently met the chief of the Russian-American Fur Company and two Russian naval officers visiting California. He also had received a letter from "Baron Wrangel, Governor of the Russian's Possessions in America and the Aleutian Islands, full of compliments, and offering me all manner of assistance, backed by Imperial favour from the court." Douglas, who had heard that Wrangel was a man "keenly alive to the interest of Science," hoped to meet the baron personally in Sitka.

While Douglas was composing a letter to Hooker, another man of science appeared in Monterey. Dr. Thomas Coulter, an Irish physician and botanist, had been serving as a medical attendant for a Mexican mining company, utilizing that position as a base for some prodigious plant collecting. Coulter shipped his specimens to a mentor in Geneva and had thought he was harvesting in untapped territory until he encountered Douglas. Instead of bristling with competition, the two men became fast friends. Coulter balanced zeal for his work with a great love of sport: as a salmon fisherman, he seemed superior even to Izaak Walton, and as a marksman with a rifle, he proved almost equal to Douglas himself. "I *do* assure you, from my heart," Douglas wrote to Hooker, "it is a *terrible pleasure* to me thus to meet a really good man, and one with whom I can talk of plants."

Adrift

No ship bound for Fort Vancouver called at Monterey over the winter of 1831, which gave Douglas the opportunity to spend another rainy season in California. "I continued to consider California still new to me," he wrote, "and to work I went a second time, finding new plants and drying better specimens." His second round of collecting around Monterey Bay produced about a hundred and fifty additional species, including another unrecorded pine, a new delphinium, and yet another mariposa lily. Then, early in 1832, political unrest in Mexico raised fears of an armed revolt. To help keep order, William Hartnell recruited Coulter and Douglas, along with fifty other foreign residents of Monterey, to enlist in La Compañía Extranjera—The Company of Foreigners. For the next two months, this volunteer peacekeeping force shared garrison duty at the presidio and remained on alert for disturbances, until new governor Don José Figueroa arrived with a contingent of soldiers to assume command. After the political situation stabilized and the rainy season ended, Coulter departed south on El Camino Real, bound for the southwestern deserts.

Although tempted to accompany his friend, Douglas did not want to miss the next ship, and so stuck close to the coast. On a short collecting trip into the Santa Lucia Mountains, he found the bristlecone fir—"you will begin to think that I manufacture pines at my pleasure," he joked to Hooker. Douglas had also received another letter from Baron Wrangel, which he quoted. "I am delighted to learn of your intended journey to our region," the baron effused. "Let me assure you, sir, that never has a visit given me more pleasure, and that you will be received with open arms in Sitka." The Russian assured him that the offer of hospitality would "of course hold good for ensuing years."

But Douglas was not able to find a ship bound for the Columbia, much less Sitka. Spring and summer both passed before he decided to sail for Honolulu, thinking that he might have a better chance there of catching a ship back to Fort Vancouver. Although he professed himself eager to leave California, a letter to George Barnston indicates that he had enjoyed his two years there well enough. The wine was "excellent, indeed, that word is too small for it; it is very excellent." He was even more effusive about the women he had met.

> The ladies are handsome, of a dark olive brunette, with good teeth, and the dark fine eyes, which bespeaks the descendent of Castille, Catalan or Leon. They (sweet creatures) have a greater recommendation than personal attractions. They are very amiable. On this head I must say, Finis, otherwise you will be apt to think, if ever I had a kind feeling for man's better half, I left it in (Calida Fornax) California.

Aside from the intimations in this letter to Barnston, Douglas did not commit to paper any details of romantic relationships in Calida Fornax (literally "hot furnace," sometimes given as the origin of the name California).

In mid-August 1832, Douglas, William Johnson, and Billy the terrier boarded an American packet that made a fast passage of nineteen days to Oahu, where British Consul Robert Charlton and his wife extended their hospitality to the collector. His gamble at finding transportation to the Columbia paid off, for the Charltons were also hosting Duncan Finlayson, a Hudson's Bay Company trader who had recently purchased a brig for the company's coastal trade and would be sailing within a few days.

In his limited time on the island, Douglas became fascinated by the juxtaposition of tropical verdure with deadly volcanic gases, and immersed himself in studying the geology of the Hawaiian chain. "In no place on the globe can the geologist better devote his time to reconcile and render harmonious this obscure but beautiful part of this exalted science, than at these islands. All that my foeble capacity can do is but a bubble."

He emerged from his bubble long enough to collect some interesting cockroaches and a pair of live *nene*, the endemic Hawaiian goose. At the Charltons', he met the captain of a South Seas whaler who offered to carry the geese to London, along with "nineteen large bundles of plants, in two chests, together with seeds, specimens of timber, etc. The Captain, a worthy little man, placed these articles in his own cabin, which gives great relief to my mind as to their safety." The *nene* geese survived the voyage around the Horn and proved to be a hit with the Zoological Society.

During his stopover in Honolulu, Douglas received a letter from Hooker conveying the news that Joseph Sabine had resigned as secretary of the Horticultural Society in response to accusations of grievous mismanagement. Rumblings against Sabine had been brewing while Douglas was in London, so this development could not have come as a complete surprise. Although Sabine and Douglas had certainly had their differences over the years, the collector felt a certain loyalty to the man who had guided his North American career. "Whatever his neglects . . . may have been, they are one hundred times covered in utter darkness itself by the good he has done in the Horticultural Society," Douglas wrote.

On September 9, 1832, Douglas wrote a letter tendering his own resignation as collector for the Society. George Bentham, who succeeded Sabine as secretary, maintained that Douglas's action resulted from a "misconception" of the circumstances. Possibly Douglas believed that he held his position at the behest of Sabine and would have no support without his patron. In a letter written to Hooker a short time later, he mentioned addressing a bundle of plants to be forwarded to Glasgow by way of Regent Street, adding that he stamped the sealing wax with a special seal given him by Hooker to discourage anyone else from opening it. This precaution "surely in these times is necessary," he wrote, implying that not everyone in the Society's employ could be trusted. He added the motto of Scottish warriors, "*nemo me impune lacessit*" (no one attacks me with impunity). Exactly who he felt was attacking him, or how he meant to punish them, he did not say.

Despite his resignation, he informed the council that he intended to honor his original agreement by continuing to send seeds and specimens from the Columbia. Three days later, he departed Oahu with Duncan Finlayson in the brig *Lama*. As the captain set a course for the Northwest coast, the collector and his instruments were on deck, ready to face the Big River once again.

XI.
THE CANYON
WINTER 1832–SUMMER 1833

A Winter with the Stars

As the *Lama* approached the mouth of the Columbia in early October 1832, her crew hailed the *Eagle*, returning from a trip up the coast. Archibald McDonald was a passenger on the *Eagle*, and as the two ships bobbed outside the river's entrance, waiting for the right conditions to cross the bar, he rowed across to the *Lama* for a visit. "Who was on board also but our old friend David after two years of perambulation over the Californias," wrote McDonald. "The two nights we were outside the Cape, depend upon it there was no lack of news, but to the exploits of our friend & his man Johnson every other topic gave way—Bears, Bulls, & Tigers had cause to rue the day they went there."

When he wasn't spinning yarns, Douglas visited with the *Eagle*'s captain, "my old friend Lieutenant Grave, R. N., who handed me a parcel from Soho Square" containing the two latest volumes of Hooker's *Flora*. During the next fortnight,

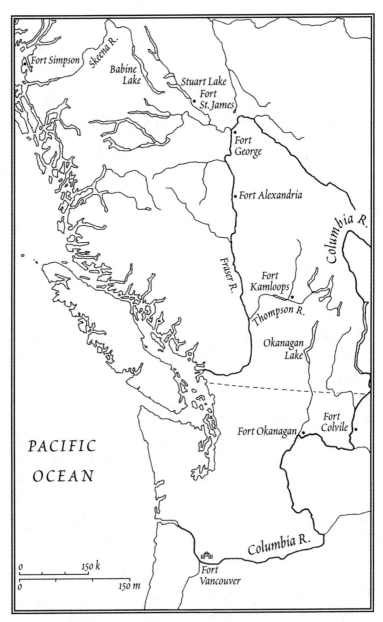

The Columbia and New Caledonia Districts of the Hudson's Bay Company, 1833

while the *Lama* remained anchored off Fort George so her crew could help load furs on the *Eagle*, Douglas revisited the sites where he had plucked so many of the plants described in those pages.

> *I carry your letter about in my note-book, and when on my walks by the side of some solitary creek, the idea not unfrequently occurs to me, that I may have overlooked some part of it, out comes your epistle for another perusal. Letters are indeed rare things to me in this part of the world.*

As the *Eagle* prepared to sail for London, Douglas sat up until 3 a.m. penning letters to Hooker and young William, whom he advised to tie some flies with the collar feathers from a blue heron, "and then you will take Trout like magic." He related a Hawaiian practice of domesticating mullet, and shared a story of cockroaches who tried to eat his shoes. "I trust we shall yet have a fine jaunt to the Highlands together, say in the summer of 1835," he closed.

In a less lighthearted letter to the professor, he stated his resolve to proceed with his original plans unless he received new orders from the Horticultural Society. He hoped to remedy the "blank in the department of sea-weeds" for the next volume of the *Flora* by writing to traders who might be visiting the coast and requesting them to gather "everything in the shape of sea-weed . . . simply coiled up, dried and put in a bag." The algae could be revived once they reached Scotland, he assured Hooker.

Regarding other gifts, Douglas had promised to secure zoological specimens for their mutual friend John Scouler and was shipping him a "tolerable collection of bones" that included a sea otter, wolves, foxes, deer, and the head of a cougar. In a touch only a true collector could appreciate,

Douglas added: "As I thought he would himself enjoy the job of cleansing them, I have only cut away the more fleshy parts, by which means, too, they hang better together."

Scouler had enraged the tribes on the lower Columbia during his 1825 visit by robbing graves "in the name of science," and Douglas had made it clear that he would not collect any human skulls for the surgeon, "lest he should be on a second voyage to the North West coast and find mine bleached in some canoe, 'because I stole from the dead,' as my old friends on the Columbia would say." Douglas added a somber note describing the fate of many of those old friends: "A dreadful intermittent fever having depopulated the neighborhood of the river, not twelve grown-up persons remain of those whom we saw when he and I were here together in 1825." One of the deceased was Chief Comcomly, who had helped introduce Douglas and Scouler to the landscape around the river's estuary.

After dispatching his mail on to the *Eagle* at the end of October, Douglas journeyed upstream to Fort Vancouver, where New England trader Nathaniel Wyeth had just arrived after a journey overland from Boston with a small party to assess the possibilities for a salmon export business. During his month-long stay on the river, Wyeth frequently visited the British post, "eating and drinking the good things to be had there and enjoying much the gentlemanly society of the place." He and Douglas bantered about the various types of grouse that inhabited the interior, and Wyeth sought the collector's advice on plants he should collect to carry back east for his close friend Thomas Nuttall.

Among other fresh faces at the post that fall was a group of young naval apprentices known as the Greenwich Boys. One of the lads, George Roberts, described Douglas as "a fair, florid, partially bald-headed Scotsman of medium stature, gentlemanly address" who traveled "with instruments for defining the positions of the places he visited." During the winter, Roberts sometimes helped the florid Scotsman with his astronomical observations, "or perhaps I should say he kindly intended to furbish up my school acquirements."

Douglas, who clearly believed in furbishing up through constant practice, logged six hundred sets of observations over the next three months. He observed Jupiter's satellites and the Earth's moon over and over in order to ascertain an absolute longitude for Fort Vancouver, which he was using as a benchmark meridian for his Northwest charts. Heartened by a letter of praise from Edward Sabine in response to measurements sent back in 1830, he gained ever more confidence in his geographical pursuits.

> *Captain Sabine goes so far as to say, that he can suggest to me no improvement in the manner of taking my astronomical or other observations, or in the way of recording them. . . . Capt. Sabine feels, I am sensible, too true a regard for my welfare not to point out my faults, and as this letter adverts to none, I may take it for granted, I trust, that he is well pleased with me.*

He observed a beautiful eclipse of the moon in early January 1833, and when a midmonth cold snap froze the Columbia solid, he braved the bitter weather night after night to take advantage of the "excellent opportunity" of clear skies for measuring the angular distances between Venus, Mercury, Saturn, and Mars.

Once the ice cleared from the river in February, Douglas and William Johnson embarked on a trip up the Cowlitz River to Puget Sound. Along the way they accumulated mosses and seaweeds for Hooker, birds for artist James Wilson, and more

geographic data for Edward Sabine. These "varied amuse-
ments," Douglas explained, kept his mind flexible: "In the
pursuit of any subject, however lofty, a man may become narrow minded, and in a condition little better than moral servitude, but by embracing different subjects we need not fear on
this head." With scant danger of lapsing into moral servitude,
he continued to sniff out "a blade of grass, a bird, or a rock,
before unnoticed." The bark of grand fir, for example, thrilled
him with its resinous blisters, pinkish red sap, and a cambium
whose consistency he could only compare to "the flesh of the
root of a Tongue of an Animal."

They had arranged to meet Archibald McDonald on Puget
Sound, where he was overseeing the first stages of construction
at Fort Nisqually (just north of modern Olympia). Favored by
fine weather, Douglas surveyed the headlands along the coast
and took altitudes of the most prominent Cascade peaks. On
their return trip in early March, he and Johnson were accompanied by Archie and his ten-year-old son Ranald, who later
described Douglas as "a sturdy little Scot; handsome rather;
with head and face of fine Grecian mould."

Douglas did not linger long at Fort Vancouver, for as usual
he would be hitching a ride with the spring express. This
year he was planning a trip far north into the remote New
Caledonia District. If all went well, he hoped to cross the
track of Sir Alexander Mackenzie on the Fraser River, then
"proceed northward, among the mountains, as far as I can do
so with safety, and with the prospect of effecting a return." In
preparation, he stocked up on rice and biscuits, tea, coffee,
and sugar, plus a gallon of wine. He selected a tent, a straw
hat, and an extra pair of deerskin trousers. Recalling other
footsore journeys, he purchased two extra sets of English
shoes and a dozen pairs of moccasins. For trade goods he took

on plenty of Irish roll tobacco and one fine lady's hat to complement the usual beads, buttons, and ribbon.

North by Twenty-two

On March 20, 1833, Douglas set off in one of the packet canoes along with William Johnson and Billy, "my old terrier, a most faithful and now, to judge from his long grey beard, venerable friend, who has guarded me throughout all my journies." Noting Billy's advancing age, Douglas vowed that he had a comfortable retirement planned for his dog, "whom, should I live to return, I mean certainly to pension off on four penny-worth of cat's-meat per day."

When the express paused at Fort Walla Walla on April 9, Douglas took advantage of the respite to write to Hooker, assuring his mentor that he moved forward with great optimism. He declared that he had never been in better health, except for his chronic ocular infirmities. While his left eye had "become infinitely more delicate and clear in its power of vision, the sight of my right eye is utterly gone; and under every circumstance it is to me as dark as midnight." He had purchased a pair of purple goggles to protect against snow blindness, but found that the lenses obscured the details of plants by blending all the colors together. The collector then confided his trepidations about the rugged terrain ahead and promised to send a copy of his daily log to Glasgow, so the professor "should know what I see and do on this most important journey."

After circling the Columbia's Big Bend, Douglas and Johnson disembarked at Fort Okanagan, where they joined a small cattle drive heading for the New Caledonia posts. This district was notorious for its lack of big game, and the traders had the idea of supplementing their monotonous diet of dried fish with fresh beef. Johnson, an experienced cowherd, could help the drovers if needed, and the slow pace of the animals would allow plenty of time for Douglas's varied scientific activities along the way.

On April 18, he opened a square, leather-bound journal of a hundred and fifty pages, composed of linen paper so fine that the volume measured only a quarter of an inch thick. A watermark on the second page clearly read

J. WHATMAN

TURKEY MILL

1827

He skipped the first hundred leaves of this notebook, then applied his pen to a left-hand page. In rich brown ink, he sketched a map that accurately depicted the serpentine Columbia receiving the Okanogan River. A neat square represented the fur trade post. A table of coordinates in the lower left corner of the page corresponded to numbered locations along the lower Okanogan River. The right leaf remained blank. Over the next month, he drew twenty-one similar maps, always on left-hand pages, documenting the cattle brigade's journey north. He calculated the coordinates of each day's resting place, designated by small Vs, recorded the mileage covered, and jotted topographical comments around his bird's-eye sketches.

From Fort Okanagan, the party trekked up the east side of the river; when a long oxbow bent west, the trail veered uphill through sand dunes to gain a high bench beneath "Rocky Hills with stunted pine trees." Looking south, Douglas calculated the height of a ridge that rose twelve hundred feet above the Columbia's long loop, then turned his attention west to view "Broken undulating ground." After a couple of hours' travel, the party would have skirted an alkaline body of water known today as Little Soap Lake, which Douglas marked with a perfect circle of good proportion. But no note survives about the soapy green water and white granite boulders that give the lake such a distinctive look, or the toxic horsebrush that crawls in spiny clumps along the rocks.

The brigade next passed Soap Lake proper, its hanging-bag shape also faithfully depicted. The water of this lake carries an even more striking emerald color, and an adjacent wetland attracts shorebirds in need of rest during their spring migration. In his field journal, Douglas might have bragged about how many dowitchers he shot, or reveled in the blue legs of an avocet probing for small crustaceans. Beyond Soap Lake, the trail bent back to the Okanogan valley and continued up the river's east side. The voyageurs staked their horses for the night just south of Mission Creek and the modern town of Omak; Douglas tallied twenty-five miles covered during the day, and "High Mountains" rising to the west. The longitude he determined as 119° 28' West—a very accurate position for one worked out during an evening camp.

Above Mission Creek, the collector climbed a "High Rocks Peak" to take a meridian altitude, but incorrectly recorded the resulting latitude as 37° rather than 47° North. Although snowy Cascade peaks stand clearly visible to the west, he focused instead on shading the river benches with endless fine lines and portraying distinctive high rocks along the valley rim. Approaching the future international border on April 20, he transcribed a correct latitude of 48° 58' 53" at the south end of Osoyoos Lake. Around that day's route, he curled phrases such as "Fine valleys with a healthy sward of grass," "Fine waterfall 200 feet high," "Grass Hills," and "Mountains covered with Pine."

Over the next three days he traced the route north of Osoyoos Lake, depicting McIntyre Bluff, a spectacular glacier-scraped wall of rock, with one spiked circle. The drive would then have wound past a showy series of falls on the main river channel before walking up the long ribbon of Okanagan Lake to its north end. From there, the Salmon River links the Okanogan and Thompson drainages, and on April 26 the cattle would have been munching through open ponderosa pine woodlands lush with emerging bunchgrass. In late April, these hills bring forth an abundance of spring beauties or Indian potatoes, an early-season tribal food that Douglas would surely have sampled.

The trail broke over the divide to Monte Lake, then followed the creek of that name into the Thompson River Valley—a glacially plowed corridor that runs west to the Fraser's main stem. On his tenth map, Douglas planted a neat square at the junction of the Thompson and North Thompson rivers to mark Fort Kamloops, where his former host from Fort Walla Walla, Samuel Black, was the agent in charge.

A strange report about Douglas's brief stop at Fort Kamloops drifted back to Fort Vancouver, where it was heard by his former helper, Greenwich boy George Roberts.

On a visit of David Douglas to Saml Blacks post at the junction of the Fraser & Thompson, while enjoying the lonely hospitality of his brother Scot, Douglas a rather fiery chap remarked the Companys Officers hadn't a soul above a beaver skin & dire was the offense, they retired with the idea of having it settled in the morning—at the early dawn of which Black was tapping at the pierced parchment which served as a window to the hut, with the query, Mr Douglas are ye ready?

This bit of lore, included in a rambling letter written forty-five years after the fact by a man who was eight hundred miles downstream when the incident took place, has been variously interpreted. Although Black and Douglas had apparently enjoyed a cordial relationship during Douglas's visits to Fort Walla Walla in 1826, both men had a history of volatility.

Black, known to be quick to take insult and slow to forgive, was reputedly so suspicious that he kept a loaded pistol under his tablecloth at meals, and he may have been one of the "enemies" that the collector had supposedly made during his first visit to the Columbia. If the two did fall into an argument over the character of the Bay Company traders, Black may have been incited to challenge his imprudent visitor to a duel. And Douglas, who more than once had expressed his quarrelsome tendency with a show of force, may have responded in kind.

George Roberts related at least two versions of the story, in one of which he declared "the man of flowers declined the winning invitation, and saved his life." The wry tone of Roberts's recollection suggests that the alleged altercation may not have been as serious as some supposed. Given the teasing hyperbole that characterized fur trade humor, Black's "challenge" could also have been a jest, but no other known account exists to help clarify the incident.

Whatever actually transpired at Kamloops, three days later, on May 1, Douglas was sketching the eleventh map in his survey book. He shaded several layers of the spectacular high hills above Kamloops Lake, then traced the thread of Cherry Creek feeding in from the south. From this point the cattle drovers continued along the Thompson River, where more open ground would have provided better grazing for the animals.

Early the following morning, the herd plodded up from the deep river valley to make for Cache Creek, passing beneath a line of deeply eroded pastel cliffs composed of Eocene shale—the depositions of a large lake that covered the area fifty million years ago. The layers of the lake bed split apart easily, like fine slate, and if Douglas had paused to examine the loose talus at the foot of the cliffs, he would have noticed strikingly clear imprints of leaves and fish on many of the pieces, now known as the McAbee fossil beds.

As the brigade paralleled the Bonaparte River into a wetter muskeg country, Douglas focused on the planet Saturn and the star Arcturus to determine his latitude. Early May always brings an abundance of nesting waterfowl to the glacier-carved lakes of this area, so the hunters would have found good eating in every direction—from solitary loons and grebes to bulky white pelicans; from courting buffleheads and hooded mergansers to rafts of redheads, widgeons, and gadwalls; from resting swans to geese on the wing. Here, on his fifteenth page, Douglas covered a detailed map of Green Lake with what in this notebook passes for a story: three Latin tree names calling out lodgepole pine, spruce, and aspen; the shape of several small islands within the lake; and the evocative phrase "High ridge of snowy mountains stretching N.W. 34 miles." If a person today climbs a slight rise above Green Lake, stunning views of the Rocky Mountains do indeed appear etched against the sky.

As Douglas filled his next page with the sinuous oval of Lac de Chevaux, the voyageurs surely would have related the gruesome story behind the lake's name—some years before, as a horse brigade en route to the Fraser clopped across its frozen surface, the ice had given way, and eighteen animals had fallen through and drowned. Rather than mark the

fatal spot, Douglas designated the trail that led safely along
the lake's shore, surrounded by remarks far removed from
thoughts of ice or water.

Low wood hills—fine rich soil

Pines on the Hills—Poplar, Willow, and Birch near the streams.

Low Woody Hills

Very Large Pine Trees

The Fraser River flows on the same order of magnitude
as the Columbia, and the approach to the great river from
Williams Lake is a stunning exercise in scale. As the brigade
turned north to follow their new course, Douglas depicted
a series of islands with their corresponding rapids. He ful-
filled his prediction to Hooker on May 9 when he entered
Fort Alexandria, named for Sir Alexander Mackenzie, who
descended the Fraser in 1793. Not far from the post site, the
explorer had realized that the river was more than he and his
crew could handle. They had abandoned their canoe to follow a tribal route overland toward their goal of the Pacific,
reaching the coast at Bella Coola.

Douglas and the supply brigade had now reached 52° 50'
North of latitude. Fort Alexandria was regarded as the northern limit of dependable pasture for livestock, and fur traders traditionally left their horses here and switched to bark
canoes or plank-sided bateaux for the run to Fort St. James.
Presumably the cattle remained at Alexandria, but there is no
record of Douglas's mode of travel as he moved north through
an undulating landscape of cottonwoods, aspen, and birch.
He completed only half of his twenty-second map, up to the
junction of the Quesnel River with the Fraser, including the
notation "numerous small lakes and marshes." That page was

followed by a swirl of brief landscape comments, but his cursory sketch lacked the anchor of the river. The last twenty-five pages of the notebook remained completely blank.

Dashed to Atoms

Fort George (on the site of the modern city of Prince George) was located at the confluence of the Fraser and Nachako rivers. One agent recalled that "the situation of the post is exceedingly dreary, standing on the right bank of Frazer's River, having in front a high hill that shades the sun until late in the morning." Douglas's party did not pause there long before making their last dash up to Fort St. James, the largest post in the New Caledonia District. This trade house was built within Dak-elh Carrier Indian territory, on the shingled south shore of Stuart Lake, which stretches out of sight to the northwest. When Douglas arrived in early May, the white-crowned sparrows would have been singing their territorial songs. Western grebes might have been rasping out their mating calls from an apron of open water along the shore, but a fresh green leaf or spring blossom would have been hard to find.

While visiting St. James, the collector learned that a small expedition was about to depart to explore the Skeena River, which feeds into Chatham Sound on the Inside Passage to Alaska. Douglas would have known that the Hudson's Bay Company had recently constructed Fort Simpson in that vicinity, and that if he could reach that post, he would be in a good position to catch a coastal trading ship to Sitka, where Baron Wrangel had assured him he would be most welcome. According to Archibald McDonald, Douglas was at first much disposed to accompany the expedition, "but fearing they could not reach the sea, or any of our settlements on the coast, and would in that case lose time, and be disappointed in other projects he had in view, he did not join the party." Instead, he borrowed a birchbark canoe from the agent at Fort St. James and began the long return trip to the lower Columbia.

The party floated back downstream to Fort George during the second week of June. Here they faced the Fraser, where spring runoff can be a wrenching experience, with the character of the river changing unpredictably from slow bergs of rotting ice to a powerful muddy deluge. About twenty miles downstream from Fort George, a wooded island splits the river. The Fraser churns with an oscillating roar as it hits this island, ripping past a sand spit that extends upstream. Driftwood spars protrude from the spit like spiked armaments, announcing the gauntlet of Fort George Canyon, where red and pink gneiss form a series of rocky islands that keep the current roiling for over two miles downstream.

On June 13, Douglas, William Johnson, Billy the terrier, and an unknown number of paddlers entered the canyon. "On that morning at the stony islands of Fraser's River, my canoe was dashed to atoms," the collector later reported. "I passed over the cataract and gained the shore in a whirlpool below, not however by swimming, for I was rendered helpless and the waves washed me on the rocks." After floundering in an eddy for more than an hour, he gained the strength to crawl ashore. He located Johnson and Billy, who had both survived, then was lucky enough to salvage his sketchbook of maps, a volume of astronomical observations, and most of his instruments along the shore. There was no sign of any of their food or other supplies, or of his plant specimens—about four hundred in number. But the loss that troubled him most was his "journal of occurrences, as this is what can never be replaced, even by myself."

Writing to Hooker some weeks later, Douglas described

a most disastrous day for me, on which I lost, what I may call, my all! . . . This disastrous occurrence has much broken my strength and spirits. I cannot detail to you the labor and anxiety this occasioned me both in mind and body, to say nothing of the hardships and sufferings I endured. Still I reflect with pleasure that no lives were sacrificed.

When George Barnston learned of the incident, he was sympathetic but philosophical—his friend had "met with another of those unfortunate accidents, that cannot always be avoided in small canoes, and which had already so often occurred to him on the waters of the Columbia." Douglas later echoed this sentiment: "On the whole I have been fortunate, considering the nature and extent of the country I have passed over, and the circumstances under which I traveled, my accidents have been few."

Douglas, Billy, and Johnson straggled back to Fort George, where they secured another canoe and made their way to Fort Alexandria without further incident. Switching to horseback, they rode over the divide to Fort Kamloops. No story has come to light describing how Samuel Black treated the pilgrims upon their arrival, but the party soon resumed their trudge to Fort Okanagan. Two Indian canoes with paddlers then ferried them down the Columbia, and on July 14, somewhere around Priest Rapids, they met a brigade commanded by Archibald McDonald, who leant an empathetic ear to the story of the "shipwreck."

The collector lingered for some time at Fort Walla Walla, making excursions into the Blues and endeavoring "as far as possible, to repair my losses, and set to work again." He was able to replace many of the lost plants, and even found some new ones that had escaped his notice previously. "I hope," he wrote to Hooker, "that this small collection may give you some satisfaction, as it is all I can now offer you from North-West America."

When Douglas reached Fort Vancouver in early August, he declared himself "greatly broken down, having suffered no ordinary toil." Soon afterward he was prostrated by a recurrence of intermittent fever. His treatment fell to Meredith Gairdner, a new doctor recruited to help McLoughlin combat the continuing outbreaks of fever on the lower river. Gairdner and his fellow physician William Tolmie (posted to Fort Nisqually) had both studied under Hooker at Glasgow. Both men were keen to follow up on the natural investigations begun by Douglas, but Dr. Gairdner had been too busy with his medical duties to venture far from Fort Vancouver since his arrival the previous May.

As the cases of fever slackened during the fall of 1833, Douglas also recovered his health, and he and Gairdner were soon laying plans to climb the Cascade's great peaks. At first they thought about ascending Mount St. Helens, but steam from an active volcanic phase discouraged that idea. The pair instead took a short excursion up the Willamette, and later made an attempt on Mount Hood. Although they did not reach the summit, Gairdner found inspiration from the excursions. "The true method of examining this country," the physician wrote, "is to follow the plan of Douglas, whether with the view of investigating the geognostic, botanical, or zoological riches of the country."

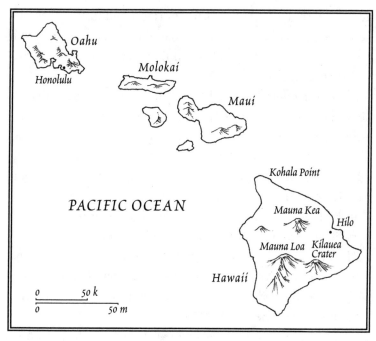

Oahu and the volcanoes of Hawaii, 1834

XII.
CRATERS
1834

Mauna Kea: White Mountain

Whether I return through the Russian Empire, or the islands of the south seas, I have not yet determined. My arrival in England is uncertain.

After his experience in the icy Fraser and the "dreary inhospitable country" of New Caledonia, Douglas apparently decided to return home by way of the South Seas rather than Siberia, for in October 1833 he and Billy boarded the *Dryad*, bound for the Sandwich Islands. As she passed the promontory of Cape Disappointment on Friday the 18th, he began a new journal notebook. Three weeks later, after a tempestuous passage, the vessel stopped in San Francisco Bay to take on food and water and make repairs, for she was, according to Douglas, "nearly a wreck."

The master of the dip needle pitched his tent on Yerba Buena Hill (now Telegraph Hill) and set up a station for astronomical observations. He took day trips to the nearby Franciscan missions and crossed the bay to Whaler's Harbor (Sausalito), noting that the ferryman's farmhouse "was placed on the site of an old Indian camp, where small mounds of

229

marine shells bespeak the former existence of numerous aboriginal tribes." Before departing on November 29, he prepared a list of latitudes and longitudes for all the nearby missions, written in careful Spanish, for Governor Figueroa.

The *Dryad* reached Oahu on December 23, and Douglas spent Christmas Day charming the wife and sister-in-law of British consul Richard Charlton in Honolulu. He had set his sights on collecting data from the big volcanoes on Hawaii, the largest of the Sandwich Islands, and Charlton suggested he contact an American missionary there. Taking passage in a light American schooner, Douglas sailed to the town of Hilo, where he called on the Reverend Joseph Goodrich, an avid mountaineer who had ascended Hawaii's Mauna Kea (White Mountain) more than once.

Goodrich was also an amateur geologist, and he and Douglas may well have discussed the concepts laid out in Charles Lyell's landmark *Principles of Geology*. Published to great acclaim three years before, Lyell's book raised thought-provoking questions about the formation of the Earth, and many Christian naturalists were trying to reconcile the creation stories from Genesis with the rocks that lay before them. The new acquaintances may have pored over a plate in Lyell's volume that depicted a cross-section of volcanic plumbing, with magma bubbling up from deep within the earth.

Since the white-topped Mauna Kea was considered the highest peak on the islands, Douglas was determined to take measurements from the vantage point of its summit. Goodrich advised him on the twenty-seven-mile ascent, then helped him line up a team of sixteen porters, explaining that eleven men would be required to pack food for the five who carried the gear. The reverend also recommended as guide and interpreter a native Hawaiian named John Honorii, a devout convert to Christianity who had visited the United States to request missionaries for Hawaii.

The party set out on January 7, 1834, with Douglas lugging a sixty-pound pack containing his key instruments. The first part of the route wound through rich lowland forests festooned with creepers and epiphytes, "while the Tree Ferns gave a peculiar character to the whole country." The landscape was far from virgin ground, however; bananas, sugarcane, coffee, and breadfruit had already been introduced as commercial crops, and feral sheep, goats, and pigs had taken their toll on the native plants. The higher scrub country abounded with wild cattle, supposedly descended from animals left behind by Captain George Vancouver during his survey of the early 1790s. Local islanders still remembered the expedition's surgeon and naturalist, Archibald Menzies, as "the red-faced man, who cut off the limbs of men, and gathered grass." Douglas reveled in the notion that he was following once again in the esteemed doctor's footsteps.

As the climbers gained elevation, steady rains began to pour down. During one torrential shower, Douglas took shelter at a sawmill owned by two Americans who greeted him hospitably, then killed a wild bullock to feed his party. While his porters cached some of this bounty along the trail for later use, the collector bounced from exotic bromeliads and silversword plants to more familiar raspberries and strawberries. As always, he paid close attention to what would best survive in an English garden, so that mosses and ferns made up the bulk of his preservations on this ascent.

They spent the next evening at a cattle hunter's rustic lodge, and Douglas took advantage of a clearing sky to pick out the constellations of Orion and Canis Minor. To his great satisfaction, he was able to focus on Canopus, one of his favorite stars—part of the keel of the mythic vessel *Argo*, and the brightest point in a vast section of tropical sky.

On the morning of January 12, Douglas and Honorii started for the summit of Mauna Kea with three porters and two Americans. The collector penned vivid descriptions of the lava forms, many now technically defined by Hawaiian terms such as *aa* and *pahoehoe*. "The general appearance is that of the channel of an immense river, heaped up," he observed. "In some places the lava seems to have run like a stream."

Walking proved difficult on the sharp talus, and the temperature dropped precipitously as they gained elevation. Douglas responded with a typical burst of energy: "Though exhausted with fatigue before leaving the Table Land, and much tired with the increasing cold, yet such was my ardent desire to reach the top, that the last portion of the way seemed the easiest." On the summit, he removed his shoes and strutted about in his stocking feet, taking measurements and collecting rocks. He was staring across the island at the snow-capped peak of Mauna Loa when he realized he had worn holes in his stockings, and his feet were paying the price. At the same time he noticed that the skin was peeling from his face, and that he had contracted a "violent headache, and my eyes became blood-shot, accompanied with stiffness in their lids."

But his physical discomforts were dwarfed by awe. "Man feels himself as nothing—as if standing on the verge of another world," he later wrote. This feeling of infinite solitude, he declared, impressed on a traveler "the extreme helplessness of his condition . . . utterly unworthy to stand in the presence of a great and good, and wise and holy God, and to contemplate the diversified works of His hands." Such religious contemplation was not new to Douglas, but perhaps due to his recent conversations with the Reverend Goodrich, his reflections on the mountain seemed more introspective. In a letter to Barnston a year before, Douglas had sworn that writing was not his game, but in his Hawaiian journal he explored facets of natural science, human history, and social dynamics with resonant prose.

With night falling and threatening clouds approaching, Honorii led Douglas and his porters back to the hunter's lodge just before a particularly fierce storm swirled around the peak. After huddling inside for two days of constant rain, they took advantage of a break in the weather to continue their descent. Slurping downhill in a sea of mud, the collector still managed to gather nearly fifty species of ferns, lichens, and mosses, including one that reminded him of a favorite green spleenwort from Scotland. Meanwhile, John Honorii became separated from the rest of the party and slipped into a swollen stream, where he clung to a rock for hours before being rescued.

Back in Hilo on January 16, Douglas compared notes with Goodrich. While Douglas had been making scientific observations on the mountain, the reverend had taken simultaneous barometric readings at sea level. After interpolating their results, they calculated the height of Mauna Kea at 13,851 feet—much lower than previous published reports, and within seventy feet of the altitude established by modern surveyors.

Into the Fire

As Douglas unwrapped his ferns and mosses from the fresh coa bark that had kept them from fermenting, then attended to the tedious business of drying and arranging them, he was already thinking ahead to Kilauea, at that time the most active volcano on the Big Island. He and John Honorii set out again on January 22, accompanied this time by only nine porters and a pared-down pile of luggage. Douglas supplied each man, including himself, with a sugarcane walking stick, "which shortens with the distance, for the pedestrian, when tired and thirsty, sits down and bites off an inch or two from the end of his staff."

The party halted for the night at a Hawaiian village, where the residents shared breadfruit and taro, squash and pumpkin, mullet and ava fish. One family was fattening a dog on the pounded *poi* of taro root; another called a black pig from the woods and slaughtered it for the visitors. While Billy fussed with two housecats over some fish entrails, Douglas only had eyes for Kilauea. Staring at "the dense black cloud which overhangs the great volcano, amid the otherwise unsullied purity of the sky," he pondered "the mighty operations at present going on in that immense laboratory."

The elevation of Kilauea is only about four thousand feet, and by the next afternoon the party had made camp near the western edge of its summit crater. Douglas sat on the lip "not correctly speaking, to enjoy, but to gaze with wonder and amazement on this terrific sight, which inspired the beholder with a fearful pleasure." Probing a nearby vent with his thermometer, he recorded a steady 158 degrees ten feet from the surface of the fissure, and 195.5 degrees directly in the crack. He and Honorii wrapped pork and a chicken in banana leaves with taro and sweet potatoes, then placed them above the fissure to cook. Steam from the Earth's core baked their meal to perfection in twenty-seven minutes.

Descending into the crater, Douglas compared the cooling lava to the breakup of ice on a northern lake, and estimated its rate of flow at almost four miles an hour. He had brought shoes for his porters to wear on the hot rocks, but soon learned that the men preferred "a mat sole, made of tough leaves, and fastened round the heel and between the toes, which seemed to answer the purpose entirely well." As three of the bearers, familiar with the maze of this volcanic landscape, guided Douglas to various scenic points, he gained an appreciation for the local sandals.

> I remained for upwards of two hours in the crater, suffering all the time an intense headache, with my pulse strong and irregular, and my tongue parched, together with other symptoms of fever. The intense heat and sulphorous nature of the ground had corroded my shoes so much, that they barely protected my feet from hot lava. . . . I took some cooling medicine, and lay down early to rest.

After dozing until 2 a.m., he awoke to a clear sky and bright moon. As was his habit, he began to shuffle about with the intention of making some observations, but soon found that the evening chill was more than his overtaxed body could bear. He "sat gazing on the roaring and agitated state of the crater, where three new fires had burst out since ten o'clock the preceding evening." Watching flames lick up from the churn of molten lava, he became aware of the sound of constant coughing—Honorii suffered from asthma, and the sulfurous smoke had brought on an attack. Douglas lent his best blanket to the guide, and a porter from Oahu contributed his own wraps, but Honorii continued to writhe and cough in misery. Meanwhile, the other porter "slept with his head toward the fire, coiled up most luxuriously and neither cold, heat, nor the roaring of the volcano at all disturbed his repose."

The next morning, the collector left one of his crew in camp to care for his papers and specimens while he and the revived Honorii circled southwest to the village of Kapapala to make preparations for their ascent of the third volcano—Mauna Loa, the Long Mountain. Their route took them through a network of lava tubes that fascinated Douglas not only with their forms and structure, but also with the ways in which the locals had adapted them for livestock pens, hot-weather retreats, garden patches, and workshops for fabricating canoes and processing mulberry bark for cloth. The headman at Kapapala prepared a clean hut for the visitors, but Douglas, ever fearful of fleas, declined the offer and pitched his tent a hundred yards away. To make sure that no one was offended, he asked Honorii to explain that privacy was required for the proper use of his instruments.

The next day was Sunday, and Honorii took leave from his guide duties to preach two sermons in the local schoolhouse. Douglas listened to the first one from the comfort of his tent, then visited the schoolhouse to watch the Bible lessons. When a young boy announced the second service by blowing on a conch shell, Douglas took his seat to listen to his friend's homily, based on the third chapter of John.

On Monday morning, Douglas hired two local guides to assist Honorii on the route up Mauna Loa—a young man named Calipio and a bird hunter who was going up the mountain in search of "that particular kind of bird which furnishes the feathers of which the ancient cloaks, used by the natives of these islands, are made." Such cloaks were fashioned from the feathers of tiny and diverse honeycreepers, whose populations must have already been rapidly declining due to the habitat deprivations of European livestock as well as demands from feather gatherers. Douglas, for all his wanderings about the island, mentioned only seeing one live bird on the wing; perhaps his dim eyesight prevented him from picking any others out of the dense of vegetation.

Two women wanted to come along on the climb, but "the bird-catcher gave them such an awful account of the perils to be undergone, that both the females finally declined the attempt." On Tuesday (January 28) Douglas set out for the mountain, followed by a colorful entourage, with most of the men wearing different combinations of sailors' uniforms and native garb. Honorii brought up the rear, carrying the telescope and an umbrella. When the asthmatic guide began coughing again, Douglas left him at a base camp in a healthy sward of grass, and pressed on with the bird-catcher in the lead. Without Honorii's leadership, the porters started to lag behind, and Douglas had some trouble getting the evening's camp under control. But as darkness came on, the clouds drifted away and he was rewarded with views of Canopus and Sirius: "Never, even under a tropical sky, did I behold so many stars."

When Douglas awoke the next morning, the man carrying his provisions still had not arrived at the camp. Skipping breakfast, the collector set out for the summit with Calipio, the bird-catcher, and two porters. They passed a waterhole without replenishing their canteens, and as the temperature dropped and the rocks grew ever more rough and glassy, the two bearers and the bird-catcher fell behind. Douglas pressed ahead with Calipio and spent the afternoon circling the crater in transfixed fascination. "The spectator is lost in terror and admiration at beholding an enormous sunken pit," he wrote, explaining his surprise at not finding the cone-shaped summit he had expected. Instead, he stood beside "a lake of liquid fire, in a state of ebullition, sometimes tranquil, at other times rolling its blazing waves with furious agitation."

One lava ledge was strung with thin filaments of glassy obsidian. The sun coaxed a sweet musical sound from these threads, which sang above a background hiss that Douglas compared to a swarm of bees. In other places, the lava had hardened into "gothic arches" of terrific magnificence. On one block of lava, he spotted a single sparrow-sized bird, light gray in color, with a faint yellow beak. "This little creature was so tame as to permit me to catch it with my hand," he wrote, "when I instantly restored its liberty. I also saw a dead hawk in one of the caves."

As so often happened, the parched Douglas waited until late in the day to leave the mountain, and he and Calipio stumbled around for some time, trying to make a rough camp in the dark. Far above timberline and with no hope of a fire, the frigid temperature soon forced them to descend by the light of Orion and a rising moon. It was a tough go. The pair finally crashed into a large clump of bushes, whereupon Douglas struck a match and determined their elevation by his trusty barometer. They had overshot their camp, but now had fuel to start a fire.

> *No sooner did its warmth and light begin to diffuse themselves over my frame, than I found myself instantly seized with violent pain and inflammation in my eyes, which had been rather painful on the mountain, from the effect of the sun's rays shining on the snow; a slight discharge of blood from both eyes followed, which gave me some relief, and which proves that the attack was as much attributable to violent fatigue as any other cause.*

At dawn he found that Calipio also suffered from eye inflammation, and as they began to trudge back up the mountain in search of their mates, thirst glued their lips together. The pair dragged into base camp to find Honorii calmly preparing breakfast. He offered them a calabash of water and ice, which refreshed their spirits; after that, "a few drops of opium in the eyes afforded instant relief both to Calipio and myself." Douglas carried small amounts of opium in the form of laudanum, and although he seldom mentioned using the remedy, this might have been the "cooling medicine" he took to relieve himself after his hard day on Mauna Kea.

By the time Douglas returned to the safe haven of Reverend Goodrich's Hilo home in early February, his body was exhausted, but his mind still whirred with images of fire and rock. In a long letter to Mrs. Richard Charlton, he tried to communicate his excitement. "As I have just about an hour ago arrived from Mauna Loa, the volcano, &c.," he began breathlessly, "and having enjoyed a bath and an excellent cup of tea, with a willing pleasure I sit down to tell you the story of a traveler."

In a feverish account, Douglas related his experience inside the Kilauea crater, at the very edge of the molten lava: "The strongest man is unstrung, the most courageous heart is daunted in approaching this place. How insignificant are the works of man." Searching for metaphors to describe the lava forms, he emphasized his ignorance and knowledge in the same overheated sentence:

> I learn that I know nothing; but this much I know, that volcanoes are the irregular, secondary results of great masses of matter obeying the primary laws of atomic action—that they differ in their intensity—are interrupted in their period, and are aggravated or constrained by an endless number of causes, external and purely mechanical.

One of the principles of geology set forth by Charles Lyell held that all the processes of earth-building could be witnessed by a careful observer. Douglas expressed hope that observing those processes at work in Hawaii would provide him with glimpses of some ultimate truth. "I assure you, Madam, that these islands offer rewards to the naturalist, over all others." Calling on biblical allusions, he meditated on the greater powers of the universe, hinting at some grand synthesis that would grant him the peace of deep under-standing. "All will be reconciled, and we shall see no longer as 'through a glass, darkly,' the infinitude, the beauty, the harmony of nature. I must return to the volcano, if it is only to look—to look and admire."

In a similar vein, he described his stumble down the slopes of Mauna Loa alongside Calipio as a return to the land of a more familiar existence: "With thankfulness and joy, the beautiful constellation of Orion being my guide, I rose to descend to a climate more congenial to my nature, and the habitations of men, the land of flowers, and the melody of birds." He had visited the summits of three great volcanoes, and he had descended into a crater. "One day there, Madam," he concluded, "is worth one year of common existence."

The Features of Our Noble Friend

After his return from Mauna Loa, Douglas mailed his Hawaiian journal to his brother John, and little is known about his activities over the next five months. Apparently he spent some weeks recovering from his mountain trek in Hilo with Reverend Goodrich. In April he traveled to Oahu, where he resided with the Charltons while waiting for a ship that would take him back to England. Yet in early May, he passed up a chance for passage home to "the Land o' cakes."

Writing to Hooker from Honolulu on May 6 about his experiences on the volcanoes, he remarked, "I go immediately to Hawaii to work on those mountains." He hoped to be back in London by the following March, and was looking forward to personally presenting the professor with the latest batch of plants he had collected for him, and to consulting "on the best mode of publishing the plants of these islands." With the packet boat that would transport him to Hilo waiting to sail, he hurriedly finished the letter, as he had so many before, with fond hopes of a reunion in Scotland. "God grant me a safe return to England. How happy shall I be to see you, Mrs. Hooker, the Boys and the Girls! William and Joseph will go to Ben Lomond. I would beg to be affectionately remembered to Dr. Scouler . . . My Captain gruffly calls me on board."

Between May and July, Douglas apparently continued to travel back and forth among the islands. While visiting Honolulu, he struck up an acquaintance with John Diell, an American chaplain at the Seamen's Chapel there. Drawn by the spell of the volcanoes, Diell accepted his new friend's suggestion of a joint excursion to the Big Island. Douglas and Billy set off in early July aboard the schooner *Minerva* with Diell's family and their black manservant, John. During a brief stopover at Maui, Diell and his family decided to make a side trip to the island of Molokai. They left John on board the *Minerva* with Douglas, planning to reunite at the Reverend Goodrich's house in Hilo.

The *Minerva* next touched land on July 9 at Kohala Point on the north end of Hawaii, where Douglas, John, and Billy disembarked to begin an overland ramble to Hilo. They had a choice of two trails: one followed the rugged ups and downs of the coast, while a second wound over the shoulder of Mauna Kea. Douglas chose the longer, gentler track, perhaps because he wanted to view the mountain from another angle before ascending it again. The distance to Hilo via the mountain trail was around ninety miles—no problem for an experienced walker like Douglas to cover in three or four days. John, however, must not have been an experienced walker, and when he pulled up lame the first evening, Douglas continued on, his few necessities tied in a small bundle, with Billy as his sole companion.

This part of Hawaii had been cattle country for some years, and Douglas and Billy spent the night of July 11 with a rancher named Davis. Early the next morning, the collector followed the path along the mountain slope until he reached an intersection of trails near the home of Ned Gurney, an English cattle hunter who supported himself by capturing the wild livestock that ranged the hillsides. He and hunters like him dug deep pits, usually around water holes, then covered the tops with light brush. When the animals came to drink, one or two would plunge into the trap, where they were easily dispatched. Like many of the island emigrants, Gurney had a checkered past. As a teenager in England, he had been convicted of stealing and deported to Botany Bay in Australia. Within three years he had slipped away from the penal colony and made it to Honolulu, and for the past decade he had scratched out a living on the Big Island.

According to Gurney, Douglas appeared at his lodge shortly before 6 a.m. on July 12. The hide-skinner provided him with breakfast, then escorted him a mile or so down the trail. Before parting, Gurney "warned him particularly of three bullock-traps, about two miles and a half ahead, near a pond." Two of them lay directly in the road, the other off to one side. Douglas, expressing a desire to reach Hilo that evening, assured his host that he could find his way and set off.

Douglas soon reached the pond surrounded by the cattle traps. He paused at the first pit, which was still camouflaged with brush. A little further along, he passed the second, where a cow had broken through the covering and fallen into the hole. He apparently continued down the path a short distance, then laid his bundle on the ground, left Billy to guard it, and turned back toward the pond.

Gurney later recounted that after setting his visitor on his way, he returned to his house and was busy skinning a bullock when two Hawaiians rushed up. They had been passing along the Hilo trail, they told him, and had paused to rest by the pond. Noticing a piece of clothing by one of the pits, they peered inside. There they saw a trapped bull, a mound of fresh earth, and, to their shock, the outline of a human foot and shoulder protruding from the dirt. They immediately rushed to Gurney for help, and upon hearing their report, he "ran into the house for a musket . . . and on arriving at the pit, found the bullock standing upon poor Douglas's body." The hide-skinner shot the bull, then climbed into the trap and extricated the corpse with the help of the two Hawaiians. He found a walking cane that his early morning visitor had been carrying, but there was no sign of the little terrier. "After a few moments' search," Gurney recalled, "the dog was heard to bark, at a little distance ahead on the road," where he sat patiently guarding Douglas's small bundle of belongings.

Gurney wrapped the collector's mangled body in a bullock hide and engaged the two natives to help him carry it to the nearest coastal village. There he hired a canoeist to transport the corpse to Hilo as quickly as possible.

Meanwhile, chaplain John Diell had reached Hilo and was staying with Reverend Goodrich, awaiting the arrival of David Douglas for their expedition to Mauna Kea. On the morning of July 14, Diell was standing in his host's doorway when a Hawaiian man rushed up with "the dreadful intelligence,

that the body of Mr. Douglas had been found on the mountain.
. . . Never were our feelings so shocked, nor could we credit the
report," Diell wrote. He and Goodrich followed the messenger
to the beach, where a canoe held a wrapped body. "Our first
care," Diell recalled, "was to have the remains conveyed to
some suitable place, where we could take proper care of them."
After carrying the body to the nearby home of a friend, they
removed the bullock hide in which it had been conveyed. "Can
it be he? Can it be he? we each exclaimed—the answer was but
too faithfully contained in the familiar articles of dress—in the
features, and in the noble person before us. They were those of
our friend," Diell wrote.

A neighbor identified as Mr. Hall helped them wash
the body, and after examining the wounds, he voiced sus-
picions that there may have been foul play involved in the
death. He had hunted wild cattle himself, and expressed
doubt whether the gashes on Douglas's face and head could
have been inflicted by the horns of a bull. Goodrich and
Diell were confused: "The matter did not seem clear—many
parts of the story were left in obscurity." In a letter posted to
Richard Charlton on July 15, the two clergymen raised a list
of perplexing questions: Why was Douglas alone? What had
become of the manservant John? How could Douglas have
fallen into a pit that he knew was there?

They had already begun digging a grave in Goodrich's
garden when they decided that a medical examination should
be made. But that would require transporting the remains to
Honolulu. And that, in turn, would require preserving the
body. "Accordingly," reported Diell, "we had the contents of
the abdomen removed, the cavity filled with salt, and placed
in a coffin, which was then filled with salt."

The reverends and their families, along with other people from the community who had known Douglas, gathered to offer prayers and pay their respects. After the service, the coffin was taken to a cool location and enclosed in a box of brine while Diell and Goodrich searched for a vessel en route to Honolulu.

At mid-afternoon on July 16, Ned Gurney arrived in Hilo carrying Billy and Douglas's belongings, which included his watch, chronometer, pocket compass, trunk keys, and a small amount of money. He gave the reverends a detailed account of all that he had seen, and sketched a map of the trail and pond, showing the locations of the three pit traps and the spot where he had found the dog. The ground had been soft enough to reveal clear imprints of Douglas's move-ments, Gurney said. After carefully scrutinizing the tracks in the area, he concluded that Douglas had for some reason laid down his bundle and turned back to view the third trap, "and whilst looking in, by making a false step, or some other fatal accident, he fell into the power of the infuriated animal, which speedily executed the work of death."

After listening to Gurney's story, Diell and Goodrich wrote to Charlton that "this narrative clears up many of the difficulties which rested upon the whole affair, and perhaps affords a satisfactory account of the manner in which Mr. D. met with his awful death."

On August 3, Douglas's remains reached Honolulu aboard a vessel from Hilo. "On opening the coffin," Richard Charlton found that "the features of our poor friend were easily traced, but mangled in a shocking manner." Charlton immediately had the body examined by two local doctors and a pair of surgeons from H.M.S. *Challenger*, then anchored in the bay. They all gave as their opinion that the wounds had been inflicted by the bullock.

Shakily satisfied with the results of these examinations, Charlton arranged to hold Douglas's funeral the next day, August 4, 1834. One of the officers of the *Challenger* read the service, and many of the city's foreign residents gathered for the interment in the cemetery of a small church. Charlton ordered that the grave be covered with brick, and suggested that Douglas's friends in England might send a stone with an inscription to mark it permanently.

Most of Douglas's personal effects, including his chronometers, reflecting circle, and dip needle, were shipped back to England, addressed to George Bentham, secretary of the Horticultural Society. Billy sailed back with the instruments, and was adopted by a clerk in the British Foreign Office.

The Doctor's Pit

Not everyone was satisfied with the conclusions of Charlton's hasty inquest. The rancher Davis, with whom Douglas had lodged the night before his death, reported that the collector had been carrying a large amount of cash—much more than was delivered to Hilo by Ned Gurney. Other possible scenarios spread quickly through the community. Some of the foreign residents suggested that the two Hawaiians who discovered Douglas's body could have mugged him and tossed him into the pit trap to cover their crime. The Hawaiian community, on the other hand, whispered among themselves that the collector had been murdered by Gurney.

Charles Hall, the cattle hunter who had helped prepare Douglas's body, remained convinced that the gashes on his face had not been made by a bull's horns. He visited the pit traps in search of clues, found the bull's body, severed its head, and shipped it to Honolulu as evidence of foul play. Two other unnamed men also journeyed to the death scene; they decided that the damage could have been inflicted by the bull's sharp hooves, and found nothing to contradict Gurney's story.

Though Douglas's tenure in Hawaii had been brief, the circumstances of his death remained a subject of continuing interest. The trap on the Hilo trail near Ned Gurney's house became a shrine of sorts, commemorated as The Doctor's Pit. Meredith Gairdner, the young physician who had treated Douglas's fever at Fort Vancouver, visited Honolulu in fall 1835 and met residents and consulate members who did not believe that Douglas "came to his end by mere accident." They maintained that it would have been impossible for anyone to fall into the pit in broad daylight. They found it curious that the chronometer Douglas carried in his breast pocket was smashed to bits, but its case was uninjured. They noted that his watch was still ticking even though its fob was torn to shreds. They repeated stories that "D. had latterly become very irritable and had frequent quarrels with the natives." As Douglas walked the Hilo trail alone, their reasoning went, some of the porters whom he had offended would have had the opportunity to harm him.

The United States Exploring Expedition visited the islands in 1841, just as Douglas's letter describing Hawaii's volcanoes to Mrs. Charlton was published in the *Hawaiian Spectator*. The expedition's geologist, the eminent James Dana, poked a little fun at the inconsistencies in Douglas's description of the craters, suggesting that the collector was befuddled from eating too much poi and roast pig. Naturalists Charles Pickering and J. D. Brackenridge, well aware of

Douglas's contributions to their trade, walked the thirty miles from Hilo to visit the pond and its three pit traps. Gurney's reputation still cast a cloud over the incident, but after examining the site, both naturalists agreed that the tragedy must have been an accident. They relayed the fate of a local cattle hunter who had dug a pit, captured a bull, then in the excitement of surveying his prize had slipped into his own trap, where he died wrestling with his captive.

Over the succeeding years, most investigators who tried to sort out this strange case have concluded that Ned Gurney probably told the truth. No one will ever know why Douglas left Billy to guard his bundle and turned back to survey the third pit. Perhaps he wanted a close look at a trapped bullock. Perhaps he had spotted an unfamiliar blossom among the brambles that covered the hole, or was stretching to grasp a wild raspberry cane. A. G. Harvey, the British Columbia historian who assembled Douglas's biographical material in the 1930s, believed that, true to form, the collector had been so confident of his abilities that he had not hired a guide who knew the area; so inquisitive that he could not resist a look into the third pit; so curious that he leaned forward a bit too far.

Irrecoverable

In late February of 1835, Archibald McDonald was on furlough in Great Britain when the first account of Douglas's death reached London. "The manner in which that melancholy event was said to have taken place seemed to us all about the Hudson's Bay House so very improbable that we were unwilling to give the report credence," he wrote.

The next day, John Douglas was reading the *Liverpool Mercury* when he happened upon a brief notice of his brother's death "while on an excursion in the mountains of Hawaii, by a wild bull." John immediately took a coach to Glasgow to inform William Jackson Hooker, who along with Edward Sabine had been planning a celebration to mark Douglas's return. John later stated "I cannot describe the state of my feelings upon reading it. Nothing ever so completely disconcerted me." Condolences arrived from noblemen and scientists, expressing "in the highest terms my late Brother's merit and exertions in the cause of science. . . . This certainly is great consolation, but my loss is irrecoverable." Before John's letter informing the rest of the family reached his mother and two sisters, William Beattie, still superintending the gardens at Scone Palace, received word from London and delivered the tragic news about his favorite pupil in person.

Douglas's childhood friend William Booth grieved that "we were in expectation of his return by the very ship which has brought us the tidings of his horrible death. . . . Such, we understand, has been the unfortunate destiny of our intrepid friend and countryman, at the early age of thirty-six. Having known him intimately from a boy, we feel a mournful pleasure in looking back to the many agreeable hours we have spent in his society, and deeply deplore his untimely fate."

Many in England's naturalist community remained skeptical until the Colonial Office and the Horticultural Society released more details of the incident, whereupon the several societies to which he had belonged issued glowing memorials. Hooker began preparing a biographical sketch and a selection of his friend's journals and letters, entitled "A Brief Memoir of the Life of David Douglas," which he published in 1836 in the *Botanical Magazine*. A few months later Edward Sabine paid his own extensive tribute by issuing "Observations taken on the Western Coast of North America by the late Mr. David Douglas, with a Report on his Papers."

A variety of Douglas's acquaintances from North America and Great Britain praised both his work and his character. Archibald McDonald, back on the Columbia, bemoaned the "sad catastrophe" and wrote "we all in this quarter lament the loss of poor Douglas exceedingly." His son Ranald, who as a young boy had traveled with the collector, remembered that "with all his enthusiasm in his pursuit, he was ever the most sociable, kindly and endearing of men."

George Barnston eulogized the companion whose lively society he had so enjoyed, now "cut down in the prime of life, and in the midst of his usefulness." Although "his successful exertions in his sphere of labour have procured him among botanists, an undying fame," Barnston mourned what might have been: "Had he lived, he would have attained to the highest celebrity as a traveller, for his diligence in investigating, and accuracy in observing, would have tended to elucidate much that is of great interest in the physical geography of the earth."

Epilogue
Nourishment beyond Names

I've always liked the desert wildflower called giant blazing-star. It's not much to look at during spring—a scraggly, rough-leafed plant with sprawling branches that crawls across the weediest of gravel slopes and crumbling trails. But as summer's heat comes on, five sharp yellow petals appear, floating above five skinny yellow sepals to form a complex star that twinkles with several dozen bright lemon stamens, fine as filaments. Many more blooms open successively atop the clump, each one as wide as your palm. On the Columbia Plateau, these stars unfold on searing afternoons, long after most other plants have gone to seed. As dusk cools the land, the stars curl up like fists and wink out.

David Douglas found giant blazing-star while walking the sun-baked Dalles portage in the summer of 1825. After carefully pressing multiple specimens, he returned that fall to shake out plenty of the distinctive winged seeds. By

1827 several plants were blooming in the Chiswick gardens; William Jackson Hooker etched an engraving for his *Flora Boreali-Americana* and noted its habit of opening in the mid-day heat. Douglas's description of the plant in Hooker's *Flora* formally named the species, and in what might have been a display of ironic humor, he devised a Latin designation that emphasized the way the stiff gray hairs of blazing-star's foliage make its rough stem appear silvery smooth: *laevicaulis*.

Such names were certainly important to Douglas. Beyond the prestige they imparted to the giver and their acknowledgement of his friends and patrons, each one integrated a new cell into the body of scientific knowledge to which he had dedicated his life. In spite of the many taxonomic alterations over the past two centuries, if you count posterity solely by titles maintained, his score remains impressive—the *douglasii* name is still attached to over eighty species of flora and fauna. That total ignores the several hundred plants such as giant blazing-star that he helped introduce into gardens, first in Great Britain, then around the world. Neither do those Latin names touch his contributions to silviculture, where his seeds of Douglas fir, Sitka spruce, and Monterey pine provided the genesis for the tallest ornamentals in Great Britain, for the most important timber-producing species in Europe, and for vast pine plantations in Britain's former colonies in the Southern Hemisphere. Such long-term horticultural and economic impacts, played against Douglas's meteoric career and mystifying demise, have kept his story alive for people who never fret about formal names.

Late June is high summer at the Dalles, windy, hot, and dry. As Douglas discovered during his journeys, however, a traveler can chase spring's freshness by following the

Columbia east to the Walla Walla River, then ascending the Blue Mountains. One rough trail leading up the open face of Blalock Mountain slices through some of the very habitats he explored, and anyone with a mind to follow David Douglas soon realizes what a good set of legs he must have had. Walker's legs, but climber's legs as well, working along mule deer paths through sliding scree and over the furrowed faces of outcrops, each step tramping through the plants that inspired him in June's profusion.

A swift wind swirls upslope to comb an ocean of sulfur lupines, pushing bright yellow waves against bunchgrass tussocks and rusty red basalt. This is the only yellow lupine Douglas encountered among well over a dozen white, blue, and purple relatives, seemingly a new species for each change of soil and exposure throughout the region. Beyond those sulfur waves, white eyebrows of receding snow draw the walker toward a dark green line that marks the forested ridge. Near the top of a tough vertical climb, the lupines give way to skeins of Douglas's onion that wrap the higher slopes with solid ribbons of pink.

The wildflower display redoubles on these scabby heights. Across one open swale, sheltered from the wind by a fractured outcrop, tapering spires of giant frasera stand tall. Among the tufts of low bunchgrass at their feet creep elegant mariposa lilies, each bloom bearing three white petals lined with purple cat hairs. Along the snowline's eroding edge crouches one perfect Brown's peony, its heavy sepals nodding over petals of velvety maroon. The fleshy leaves and thick yellow stamens look as if they belong in a tropical greenhouse, but this is not the tropics. Even now a racing cloud blots out the sun to pelt the ridge with snow and hail, the same sort of squall that Douglas dodged while gathering wildflowers here. It's a moment to savor the steady flow of description and anecdote that accompanied each of his new finds, and to marvel at the ingenuity he employed to get them all back to England.

The squall moves on after assaulting other flowers Douglas would have seen: blue camas and wild hyacinth, pink bitterroot, yellow and white *Lomatiums*. This is a moment to remember that just as surely as his collections changed many gardeners' and scientists' views of the Columbia country, the Northwest changed him. Most of the fur traders with whom he traveled had emerged from Scottish roots that mirrored his own, and he saw how they adapted to their new world. He shared food and shelter with mixed-blood hunters and savored the dark sweet taste of baked camas and black lichen cakes with tribal families. He watched women and children apply their digging sticks to many of the plants he was collecting for show, listening to their stories as they worked their own exquisite gardens.

Not least among Douglas's gifts was an ability to communicate his enthusiasm to the astonishing variety of people that he met on his journeys. William Jackson Hooker's son Joseph Dalton was only one of many acquaintances who kept that flame alive. The youngster who loved to fish with Douglas grew up to become an academic botanist and the director of Kew Gardens, just like his father. But Joseph also took to the field, making far-flung collecting trips from the Himalayas to the Antarctic. After touring western America, he wrote a paper comparing the vegetation of the Rocky Mountains with that of Asia, thus tapping one of Douglas's own unfulfilled dreams.

In 1845, Joseph Dalton agreed to classify some dried specimens from Tierra del Fuego and the Galapagos Islands that had sat unexamined since Charles Darwin pressed them during his voyage on the *Beagle* ten years earlier. As Hooker began to sift through the wrinkled blotters and brittle leaves, he realized their singular importance, just as David Douglas and John Scouler had sensed during their own Galapagos visit. Those plants, and Joseph Hooker's patient manner, helped Darwin formulate his ideas for *The Origin of Species*.

David Douglas was surely one link in the long line of collecting naturalists who led to such epiphanies. He had wrestled with large questions concerning life and species while he scampered about in hot stockings in the Kilauea crater, trying to reconcile all that he had seen and learned about the natural world. He touched those same eternal mysteries as he marveled at the hydraulic force of Lake Waha's artesian spring, or pondered a new species of lupine, or recognized the outsized hooves of woodland caribou as a perfect adaptation for walking across snow, or squatted around an earth oven with teasing tribal women, anticipating the next strange root that would emerge from the coals.

The thrill of such discoveries lifted him from the sagebrush summer into Blue Mountain snow squalls, where he could disappear in a riot of June wildflowers. On the shattered basalt atop Mount Blalock, the collector knelt close to judge the progress of one season's royal peony seeds. During the course of some future excursion, when the elements had toasted them to perfection, he would gather a few. That was one of the ways Douglas would share part of this place with the rest of the world: he would take those seeds home, and encourage them to sprout anew.

ACKNOWLEDGMENTS

IN MEMORY *of Alice Ignace, who loved plants, and of Walt Goodman, Lloyd Keith, and Robert Sherwood*

GRATEFUL ACKNOWLEDGMENTS to all the people and institutions who so generously contributed to this book, in so many ways. They include Kathy Ahlenslager, Nancy Anderson, Meryl Andrew, Felix Aripa, Karen Beaudette, Barb Benner, J.R. Bluff, Steve Box, Raymond Brinkman, Angela and Rex Buck, Pam Camp, Tucker Childs, Bobbi Conner, Chalk Courchane, Francis Cullooyah, Ed Cunningham, Dennis Dauble, Pauline Flett, Orten Ford, Vi Frizzell, Darlene Garcia, Ray Gardner, Nancy Graybeal, Joan Gregory, Jim Groth, Laurel Hansen, Larry Hufford, Gene Hunn, Kate Kershner, Bill Leonard, Chris Loggers, Leonard Louie, Gary Luke, Jack McMaster, Nancy Molina, Melinda Morning Owl, Lynn Pankonin, Nicholette Prince, Bruce Rigsby, Ann Rittenberg, Don Ross, John Ross, Tim Ryan, Sasquatch Books, Gillian Sayer, John Schenk, Michael Schroeder, Sherri Schultz, Diane Stutzman, Quentin Terbasket, Bruce Watson, Mike Webster, the students and staff of Wellpinit school, Jean Wood, Marian Wynecoop, Marsha Wynecoop, Rich Zack, and Henry Zenk.

British Columbia Archives
Butler Library, Columbia University
Conner Museum and Ownbey
Herbarium, Washington State
University
Fort Okanagan Historical Park
Hudson's Bay Company Archives

Kew Gardens, London
Linnaean Society, London
New York Botanical Society
Public Record Office, London
Royal Botanic Gardens, Kew
Royal Horticultural Society, London
Spokane Public Library

And especially Claire

BIBLIOGRAPHY

Abbott, Douglas A., and Edward Sabine. "Observations Taken on the Western Coast of North America and with a Report on His Paper." *Abstracts of the Papers Printed in the Philosophical Transactions of the Royal Society of London* 3 (1830–1837): 471–72.

Allan, Mea. *The Hookers of Kew.* London: Michael Joseph Ltd., 1967.

Anderson, Margaret J. *Olla-Piska: Tales of David Douglas.* Portland: Oregon Historical Society, 2006.

Arno, Stephen F., and Ramona P. Hammerly. *Northwest Trees.* Seattle: The Mountaineers, 1977.

Balfour, F. R. S. "David Douglas." *Journal of the Royal Horticultural Society, London* 68 (1942): 121–128, 153–162.

Ball, John. *Born to Wander: Autobiography of John Ball, 1794–1884.* Grand Rapids, MI: Grand Rapids Historical Commission, 1925.

Barnston, George. "Abridged Sketch of the Life of Mr. David Douglas, Botanist, with a Few Details of His Travels and Discoveries." *Canadian Naturalist and Geologist* 5 (1860): 120–32, 200–208, 267–78, 329–49.

Bartram, William. *Travels.* Philadelphia, 1791. Reprint, edited by Mark Van Doren. New York: Dover, 1955.

Bell, Thomas. "Description of a New Species of Agama, Brought from the Columbia River by Mr. Douglass." *Transactions of the Linnean Society of London* 16 (1833): 105–7.

Bentham, George. *Autobiography 1800–1834.* Edited by Marion Filipiuk. Toronto: University of Toronto Press, 1996.

Black, Samuel. *Faithful to Their Tribe & Friends: Samuel Black's 1829 Fort Nez Perces Report.* Edited by Dennis Baird. Northwest Historical Manuscript Series. Moscow, ID: University of Idaho Library, 2000.

Booth, William Beattie. Memorandum on David Douglas. Royal Botanic Gardens, Kew, England.

_____. "Some Account of the Late Mr. Douglas." *Gardener's Magazine* 11 (1835): 271–72.

Bouchard, Randy, and Dorothy Kennedy. "Report Prepared for the B.C. Heritage Conservation Branch, Victoria." August 1985.

Boyd, Robert. *The Coming of the Spirit of Pestilence.* Vancouver: University of British Columbia Press, 1999.

_____. *Indians, Fire, and the Land in the Pacific Northwest.* Corvallis, OR: Oregon State University Press, 1999.

Brown, Janet. *Charles Darwin: A Biography.* 2 vols. New York: Knopf, 1995, 2002.

_____. *Darwin's Origin of Species: A Biography.* New York: Atlantic Monthly Books, 2006.

Brown, Jennifer S. H. *Strangers in the Blood: Fur Trade Company Families in Indian Country.* Vancouver: University of British Columbia Press, 1980.

Brown, William C. "Old Fort Okanogan and the Okanogan Trail." *Oregon Historical Society* 15 (March 1914): 1–36.

Clinton, DeWitt. Papers. Columbia University Library, New York.

Cole, Jean Murray. *This Blessed Wilderness.* Vancouver: University of British Columbia Press, 2002.

Courchane, David C. "The Descendants of James Finlay." In possession of the author.

Cottage Gardener 6 (July 31, 1851): 263–64.

Coville, Frederick V. "The Botanical Explorations of Thomas Coulter in Mexico and California." *The Botanical Gazette* (December, 1895): 519–30.

Cox, E. H. M. "The Plant Collectors Employed by the Royal Horticultural Society, 1804–1846." *Journal of the Royal Horticultural Society* 80 (1955): 264–80.

Creech, E. P. "Brigade Trails of B.C." *The Beaver* (March 1953): 10–15.

Dakin, Susanna Bryant. *The Lives of William Hartnell.* Stanford, CA: Stanford University Press, 1949.

Dana, James D. "Geology." In *United States Exploring Expedition under the Command of Charles Wilkes, U.S.N.* Vol. 10. New York: George P. Putnam, 1849.

Davies, John. *Douglas of the Forests.* Seattle: University of Washington Press, 1980.

Davis, William Heath. *Seventy-five Years in California.* San Francisco: John Howell, 1929.

Desmond, Ray. *Dictionary of British and Irish Botanists and Horticulturalists.* London: Taylor and Francis, 1994.

Douglas, David. "An Account of a New Species of *Pinus*, Native of California."

Transactions of the Linnean Society 15 (1827): 497–500.

_____. "An Account of Some New, and Little Known Species of the Genus *Ribes*." *Transactions of the Horticultural Society* 8 (1830): 509–18.

_____. "An Account of the Species of *Calochortus*; a Genus of American Plants." *Transactions of the Horticultural Society* 7 (1830): 275–80.

_____. "Description of a New Species of the Genus *Pinus (P. Sabiniana)*. *Transactions of the Linnean Society* 16 (1833): 747–50.

_____. Field Sketches, Fort Okanagan to Quesnel River, 1833. British Columbia Archives, Victoria.

_____. *Journal Kept by David Douglas During His Travels in North America.* London: William Wesley and Son, 1914.

_____. Letter to Mrs. Richard Charlton. *Hawaiian Spectator*, April 1839, pp. 98–101.

_____. "Observations of Some Species of the Genera *Tetrao* and *Ortyx*." *Transactions of the Linnean Society* 16 (1833): 133–49.

_____. "Observations on the *Vultur californianus* of Shaw." *Zoological Journal* 4 (1829): 328–30.

_____. "Observations on Two Undescribed Species of North American Mammalia." *Zoological Journal* 4 (1829): 330–32.

_____. Papers. Lindley Library. Royal Horticultural Society, London.

_____. "A Sketch of a Journey to the North-Western Parts of the continent of North America, during the Years 1824, 5, 6, and 7." *Companion to the Botanical Magazine* 2 (1836): 82–177.

_____. "Sketch of a Journey to North-West America, 1824–27." In *Journal Kept by David Douglas, 1823–1827*. London: William Wesley & Son, 1914.

_____. "Volcanoes in the Sandwich Islands." *Geographical Society Journal* 4 (1834): 333–343.

Douglas, John. Biographical Notes of David Douglas. Royal Botanic Gardens, Kew, England.

Drummond, Thomas. "Sketch of a Journey to the Rocky Mountains and to the Columbia River in North America." *Botanical Miscellany* 1 (1830): 178–219.

Drury, Clifford M. "Oregon Indians in the Red River School." *Pacific Historical Review* 7 (1938): 50–60.

Elliott, Brent. *The Royal Horticultural Society: A History, 1804–2004*. Chicester, England: Phillimore & Co., 2004.

Ermatinger, Edward. "Edward Ermatinger's York Factory Express Journal . . . 1827–28." Edited by C. O. Ermatinger and James White. *Proceedings and Transactions of the Royal Society of Canada*, 3rd ser., 6 (1912): 67–132.

Fleming, J. H. "The California Condor in Washington." *The Condor* 26 (May 1924): 111–12.

Fletcher, Harold R. *The Story of the Royal Horticultural Society, 1804–1968*. London: Oxford University Press, 1969.

Fort Vancouver Account Books. Hudson's Bay Company Archives, Winnipeg, Manitoba.

Fort Vancouver Correspondence. Hudson's Bay Company Archives, Winnipeg, Manitoba.

Franklin, John. *Narrative of a Journey to the Shores of the Polar Sea in the Years 1819-20-21-22*. London: John Murray, 1824.

Gairdner, Meredith. Correspondence. William Jackson Hooker Papers. Royal Botanic Gardens, Kew, England.

Geyer, Carl Augustus. "Notes on the Vegetation and General Character of the Missouri and Oregon Territories." *London Journal of Botany* 5 (1846): 296.

Gibson, James R. *The Lifeline of the Oregon Country: The Fraser-Columbia Brigade System, 1811–47*. Vancouver: University of British Columbia Press, 1997.

Goldie, John. "In Memory of David Douglas." *British Columbia Historical Quarterly* 2 (April 1938): 89–94.

Goode. Memorandum. Lindley Library. Royal Horticultural Society, London.

Graustein, Jeannette E. *Thomas Nuttall, Naturalist; Explorations in America, 1808–1841*. Cambridge, MA: Harvard University Press, 1967.

Gunther, Erna. *Ethnobotany of Western Washington: The Knowledge and Use of Indigenous Plants by Native Americans*. Seattle: University of Washington Press, 1973.

Hall, F. S. "Studies in the History of Ornithology in the State of Washington (1792–1932) with Special Reference to the Discovery of New Species: Part III." *The Murrelet* 15 (January 1934): 2–19.

Harrington, John Peabody. Papers. Smithsonian Anthropological Archives, Washington, D. C.

Harrison, Peter. *Seabirds of the World*. Princeton, NJ: Princeton University Press, 1987.

Harvey, Athelstan G. "Chief Concomly's Skull." *Oregon Historical Quarterly* 40 (1939): 161–67.

———. "David Douglas in British Columbia." *British Columbia Historical Quarterly* 4 (October 1940): 221–43.

———. *Douglas of the Fir: A Biography of David Douglas, Botanist.* Cambridge, MA: Harvard University Press, 1947.

———. "Meredith Gairdner: Doctor of Medicine." *British Columbia Historical Quarterly* 9 (1945): 89–112.

Hickman, James C., ed. *The Jepson Manual: Higher Plants of California.* Berkeley: University of California Press, 1993.

Hitchcock, C. Leo, and Arthur Cronquist. *Flora of the Pacific Northwest: An Illustrated Manual.* Seattle: University of Washington Press, 1973.

Hooker, William Jackson. "A Brief Memoir of the Life of Mr. David Douglas, with Extracts from His Letters." *Companion to the Botanical Magazine* 2 (1836): 79–82, 177–182.

———. *Flora Boreali-Americana; being the Botany of the Northern Parts of British America.* 3 vols. London: Henry G. Bohn, 1829–40.

———. "On the Botany of America." *Brewster's Edinburgh Journal of Science* 2 (1825): 108–29.

———. Papers. Royal Botanic Gardens, Kew, England.

Horticultural Society of London. Minutes of Council. Lindley Library, Royal Horticultural Society, London.

Howay, F. W. *British Columbia: From the Earliest Times to the Present.* Vol. 2. Vancouver, BC: S. J. Clarke Publishing Company, 1914.

Howell, John Thomas. "A Collection of Douglas's Western American Plants." *Leaflets of Western Botany* 2 (1937–39): 59–62, 74–77, 94–96, 116–19, 139–44, 170–74, 189–92.

Hudson's Bay Company Archives, Winnipeg, Manitoba.

Hunn, Eugene. *Nch'I=Wana: The Big River: Mid-Columbia Indians and Their Land.* Seattle: University of Washington Press, 1990.

———, and Thomas Morning Owl. *In a Land Called 'tiicham': A Sahaptian Language Place Names and Ethnographic Atlas.* Pendleton, OR: Confederated Tribes of the Umatilla Indian Reservation, in press.

Keyser, James D. *Indian Rock Art of the Columbia Plateau.* Seattle: University of Washington Press, 1992.

Knight, Thomas Andrew. *A Selection from the Physiological and Horticultural Papers . . . by the late Thomas Andrew Knight, Esq. . . . to which Is Prefixed a Sketch of His Life.* Edited by Frances Acton. London, 1841.

LaLande, Jeff. *First Over the Siskiyous: Peter Skene Ogden's 1826–27 Journey through the Oregon-California Borderlands.* Portland: Oregon Historical Society Press, 1987.

Lambert, Aylmer Bourke. *A Description of the Genus Pinus.* 3 vols. London: Weddell, Prospect Row, Walworth, 1803, 1828, 1832.

Layman, William D. *Native River: The Columbia Remembered.* Pullman: Washington State University Press, 2002.

Lewis, James. "Biographical Notice of the Late Thomas Nuttall." *Proceedings of the American Philosophical Society* 7 (January–June 1860): 297–315.

Lindley, John. *Edwards's Botanical Register*. Vols. 12–19. London: James Ridgeway, 1826–32.

Log of the *Willliam and Ann*, 1824–25. Hudson's Bay Company Archives, Winnipeg, Manitoba.

Lyell, Charles. *Principles of Geology*. Vol. 1. London: John Murray, 1830.

MacDonald, Ranald. *Ranald MacDonald, the Narrative of his Early Life*. Edited by William S. Lewis and Naojiro Murakami. Spokane: Eastern Washington Historical Society, 1923.

McDonald, Archibald. Papers. British Columbia Archives, Victoria.

McLeod, Alexander. Journal. Hudson's Bay Company Archives, Winnipeg, Manitoba.

McLoughlin, John. *Letters of Dr. John McLoughlin Written at Fort Vancouver 1829–32*. Edited by Burt Brown Barker. Portland, OR: Binford & Mort, 1948.

_____. *Letters of Dr. John McLoughlin Written at Fort Vancouver to the Governor and Committee*. Edited by E. E. Rich. London: The Hudson's Bay Record Society, 1941.

Markham, Violet R. *Paxton and the Bachelor Duke*. London: Hodder and Stoughton, Ltd., 1935.

Mayne, Richard C. *Four Years in British Columbia and Vancouver Island*. London: John Murray, 1862.

Meyers, J. A. "Finan McDonald—Explorer, Fur Trader, and Legislator." *Washington Historical Quarterly* 13 (1922): 196–208.

Mitchell, Ann Lindsay, and Syd House. *David Douglas: Explorer and Botanist*. London: Arum Press, 1999.

Morwood, William. *Traveler in a Vanished Landscape: The Life and Times of David Douglas*. London: Gentry Books, 1973.

Moulton, Gary, ed. *The Journals of the Lewis & Clark Expedition*. 12 vols. Lincoln: University of Nebraska Press, 1983–2001.

Nisbet, Jack. *The Mapmaker's Eye: David Thompson on the Columbia Plateau*. Pullman: Washington State University Press, 2005.

_____. *Sources of the River: Tracking David Thompson across Western North America*. Seattle: Sasquatch Books, 1994, 2007.

_____. *Visible Bones: Journeys across Time in the Columbia River Country*. Seattle: Sasquatch Books, 2003.

Nuttall, Thomas. *The Genera of North American Plants*. Philadelphia: D. Heartt, 1818.

_____. *A Journal of Travels into the Arkansas Territory during the Year 1819*. Philadelphia: Thos. H. Palmer, 1821.

Ogilvie, J. F. "A Pilgrimage up the Athabasca Pass." *Scottish Forestry* 38 (1984): 119–24.

_____. "A Portrait of David Douglas." *Arboricultural Journal* 4 (1980): 119–25.

Peattie, Donald Culross. *A Natural History of Western Trees*. New York: Bonanza Books, 1950.

Phillips-Wolley, Clive. "Mining Development in British Columbia." *The Canadian Magazine* (February 1897): 299–304.

Pursh, Frederick. *Flora Americae Septentrionalis*. London: White, Cochrane, and Co., 1814.

Reed, Henry E. "William Johnson." *Oregon Historical Quarterly* 34 (1933): 314–23.

Rich, E. E. *The Fur Trade and the Northwest to 1857.* Canadian Centenary Series. Toronto: McClelland and Steward, 1967.

———. *History of the Hudson's Bay Company,1670–1870.* 2 vols. London: Hudson's Bay Record Society, 1960.

Richardson, John. Correspondence. William Jackson Hooker Papers. Royal Botanic Gardens, Kew, England.

———. *Fauna Boreali-Americana, or the Zoology of the Northern Parts of British America.* 2 vols. London: John Murray, 1829.

Robbins, Christine Chapman. *David Hosack: Citizen of New York.* Philadelphia: American Philosophical Society, 1964.

Roberts, George B. "Letters to Mrs. F. F. Victor." *Oregon Historical Quarterly* 63 (1962): 175–205.

———. Recollections of the Hudson's Bay Company. Bancroft Library, University of California, Berkeley.

Robinson, Alfred. *Life in California during a Residence of Several Years in That Territory.* New York: 1846. Reprint, New York: Da Capo Press, 1969.

Rodgers, Andrew Denny. *John Torrey: A Story of North American Botany.* Princeton: Princeton University Press, 1942.

Ruby, Robert H., and John A. Brown. *The Spokane Indians: Children of the Sun.* Norman: University of Oklahoma Press, 1970.

Russell, Israel Cook. "Geology and Water Resources of Nez Perce County, Idaho." *Water Supply and Irrigation Papers of the U.S. Geological Survey,* no. 53 (1901): 79–81.

Sabine, Edward. "Report on the Variations of Magnetic Intensity Observed at Different Points of the Earth's Surface." *Report of the Seventh Meeting of the British Association for the Advancement of Science* 6 (1837): 1–85; 500.

———. Observation Taken on the Western Coast of North America, by the Late Mr. David Douglas with a Report on His Papers. MS-0623. Archives of British Columbia, Victoria.

Scouler, John. "Dr. John Scouler's Journal of a Voyage to N.W. America." *Oregon Historical Quarterly* 6 (1905): 54–75, 159–205, 276–87.

———. "Remarks on the Form of the Skull of the North American Indian." *Zoological Journal* 4 (1829): 304–09.

Simpson, George. Character Book. Hudson's Bay Company Archives, Winnipeg, Manitoba.

———. Correspondence. Hudson's Bay Company Archives, Winnipeg, Manitoba.

———. *Fur Trade and Empire: George Simpson's Journal, 1824–25.* Edited by Frederick Merk. Cambridge, MA: Belknap Press of Harvard University Press, 1968.

Suttles, Wayne, and William Sturtevant, eds. *Handbook of North American Indians.* Vol. 7. *Northwest Coast.* Washington, DC: Smithsonian Institution, 1998.

Sweetser, Albert R. "David Douglas, Botanist." *Sunday Oregonian,* November 7, 1926, sec. 5, p. 5.

_____. "Members of Plant Kingdom Stand as Living Memorials to Work on Pacific Coast of David Douglas, Botanist." *Sunday Oregonian*, April 12, 1925, sec. 5, p. 11.

Thorington, J. M. "The Centenary of David Douglas' Ascent of Mount Brown." *Canadian Alpine Journal* 16 (1926–27): 185–97.

Tolmie, William Fraser. *The Journals of William Fraser Tolmie, Physician and Fur Trader*. Vancouver, BC: Mitchell Press Limited, 1963.

_____. Papers. British Columbia Archives, Victoria.

Transactions of the Horticultural Society of London, 1824–34.

Turner, Nancy. *Food Plants of Interior First Peoples*. Victoria: Royal British Columbia Museum, 1997.

_____, Randy Bouchard, and Dorothy Kennedy. *Ethnobotany of the Okanagan-Colville Indians of British Columbia and Washington*. Occasional Papers of the British Columbia Provincial Museum, no. 21, 1980.

_____, Laurence C. Thompson, M. Terry Thompson, and Annie Z. York. *Thompson Ethnobotany: Knowledge and Usage of Plants by the Thompson Indians of British Columbia*. Victoria: Royal British Columbia Museum, 1990.

Vitt, Dale H., Janet E. Marsh, and Robin B. Bovey. *Mosses, Lichens, & Ferns of Northwest North America*. Vancouver, BC: Lone Pine Publishing, 1988.

Von Kirk, Sylvia. *Many Tender Ties: Women in Fur-Trade Society in Western Canada, 1670–1870*. Winnipeg, Manitoba: Watson & Dwyer Publishing, 1981.

Walker, Deward E. Jr., ed. *Handbook of North American Indians*. Vol. 12, *Plateau*. Washington, DC: Smithsonian Institution, 1998.

Watson, Bruce. Biographies of Hudson's Bay Company employees in the Columbia and New Caledonia Districts. In the author's possession.

Wheeler, Arthur O. "Mounts Brown and Hooker." *Canadian Alpine Journal* 17 (1928): 66–68.

Wilkes, Charles. *Narrative of the United States Exploring Expedition During the Years 1838, 1839, 1841, 1842*. Vol. 4. Philadelphia: Lea and Blanchard, 1849.

Wilson, James. *Illustrations of Zoology*. Edinburgh: William Blackwood, 1831.

Wilson, William F. "David Douglas, Botanist, at Hawaii." Honolulu, n.p., 1919.

Work, John. "The Journal of John Work." Edited by T. C. Elliott. *Washington Historical Quarterly* 6 (1915): 26–49.

Wyeth, Nathaniel J. "Correspondence and Journals." In *Sources of the History of Oregon*, edited by F. G. Young. Eugene: University of Oregon Press, 1899.

Zenk, Henry B., and Tony Johnson. "Uncovering the Chinookan Roots of Chinuk Wawa."*The University of British Columbia Working Papers in Linguistics* 14 (2004): 419–51.

CHAPTER NOTES

CBM: *Companion to the Botanical Magazine* 2 (1836).

DCB: *Dictionary of Canadian Biography*

HBCA: Hudson's Bay Company Archives, Winnipeg, Manitoba

Prologue
Unless otherwise noted, all quotes are from David Douglas, *Journal* (1914), 258–59.

p. xi: Climbing Mount Brown: Ogilvie, "A Pilgrimage"; Thorington, "Centenary"; Wheeler, "Mounts Brown and Hooker"; and Don Ross, personal communication.

Chapter 1
Unless otherwise noted, all quotes are from David Douglas, *Journal* (1914), 1–30.

p. 2: *nymphoides:* a species of aquatic flowering plants.

p. 3: Douglas's early life: Booth, Memorandum; *Cottage Gardener,* 263; Hooker, "Brief Memoir," CBM, 79–80.

p. 3: "his contempt for the master's thong": *Cottage Gardener,* 263.

p. 4: John Douglas's remembrances: Douglas, John; Hooker, "Brief Memoir," CBM, 79–80.

p. 5: Douglas's apprenticeship: Booth to Hooker; Douglas, John.

p. 6: "Douglas was so enthusiastic": *Cottage Gardener,* 263.

p. 6: "once rushed into the lecture room": ibid.

p. 7: "great activity, undaunted courage": Hooker, "Brief Memoir," CBM, 82.

p. 7: "an individual eminently calculated": ibid.

p. 8: Botany's economic importance: Janet Brown, *Charles Darwin,* I: 91–92.

p. 8: "distinguished persons": Markham, *Paxton,* 53.

p. 8: Horticultural Society of London: Elliott, *Royal Horticultural Society,* 1–12.

p. 9: William Atkinson's Chiswick design: ibid., 62.

p. 11: Torrey to Hooker: Graustein, *Thomas Nuttall,* 194–95.

p. 16: Person's *Synopsis Plantarum:* Christiaan Hendrick Person (1761–1836) was a Dutch botanist and mycologist who published an early two-volume manual of plants using the Linnean system.

p. 18: gardeners furious against Douglas: Richardson to Hooker, April 22, 1825, Hooker Papers.

p. 19: Nuttall's scientific education: Graustein, *Thomas Nuttall,* 4–32.

p. 21: Clinton's prize pigeons: Douglas to Clinton, October 3, 1825, Clinton Papers.

Chapter 2
Unless otherwise noted, all quotes are from David Douglas, *Journal* (1914), 77–102.

p. 23: "His mission was executed": *Transactions of the Horticultural Society of London* 5 (1824): v.

p. 24: "embraced nearly everything": ibid.

p. 24: "done himself great credit": Hooker, "On the Botany of America," 127.

p. 24: Clinton comments: Clinton to Sabine, October 10 and 28, 1823, Clinton Papers.

p. 24: "the shyest being": Knight, *A Selection*, 39–40.

p. 25: "Some Account of the American Oaks . . .": Douglas, *Journal*, 31–49.

p. 25: "find the fare of the country rather coarse": HBCA, Official Correspondence Files, June 24, 1824.

p. 27: *William and Ann* roster: Log of *William and Ann*, 1824–25, HBCA C.1/1066.

p. 27: Scouler "as an opportunity of prosecuting": Douglas, "Sketch," CBM, 83.

p. 27: Scouler as "a man skilled in several": ibid.

p. 28: "small jellyfish": Scouler, "Journal," 57.

p. 29: "my pockets filled with the granite": ibid., 61.

p. 30: Seabird identifications: Harrison, *Seabirds*.

p. 30: Scouler albatross dissection: Scouler, "Journal," 62.

p. 31: "the place we landed": ibid., 65.

p. 32: Captain Hanwell: Log of the *William and Ann*, January 10, 1825, HBCA C.1/1066.

p. 33: "The Gallipagos as will be seen": Scouler, "Journal," 74.

p. 34: "the fracture was in the middle third": ibid., 161.

p. 34: "the quick motions of the vessel": Log of the *William and Ann*, HBCA C.1/1066.

Chapter 3
Unless otherwise noted, all quotes are from David Douglas, *Journal* (1914), 101–29.

p. 38: English words: Scouler, "Journal," 163.

p. 39: Alexander McKenzie bio: Watson, Biographies.

p. 39: Modern linguists: Zenk and Johnson, "Uncovering the Chinookan Roots."

p. 39: Elk-hide shields: Scouler, "Journal," 167.

p. 40: Young horsetail stems: probably field horsetail, *Equisetum arvense*. Gunther, *Ethnobotany*, 15.

p. 40: "We have now so completely ransacked": Scouler, "Journal," 172.

p. 42: "Mr. Douglas is a Passenger": July 22, 1824, HBCA, A.6/21, fo. 11d.

p. 42: Wild raspberry shoots: probably thimbleberry, *Rubus parviflorus*. Gunther, *Ethnobotany*, 34.

p. 44: Archibald McDonald: Cole, *Blessed Wilderness*, 102–9.

p. 45: "according to the custom of the country": Jennifer Brown, *Strangers*, 1980.

p. 46: Asphodel: probably the lily, sticky tofieldia (*Tofieldia glutinosa*), which resembles the European asphodel.

p. 52: Rock art at the Dalles: Keyser, *Indian Rock Art*, 16–32.

Chapter 4
Unless otherwise noted, all quotes are from David Douglas, *Journal* (1914), 130–57.

p. 54: Pursh's appeal to botanists: Pursh, *Flora*, xii.

p. 56: Beach peavine: *Lathyrus littoralis*. Hitchcock, *Flora*, 263.

p. 56: Seaside lupine: *Lupinus littoralis*. Gunther, *Ethnobotany*, 38.

p. 56: Fern fronds as decorative garlands: Hooker, *Flora*, 2: 261.

p. 57: Finan McDonald's career: Nisbet, *Sources of the River*, 81–88.

p. 58: "elegant movements": Douglas, "Sketch," CBM, 92.

p. 58: Trade versus native tobacco: Nisbet, *Visible Bones*, 132–52.

p. 59: "fraught with treasures": Douglas, "Sketch," CBM, 92.

p. 60–61: Drainages on the south side of Cascade Rapids: Nancy Molina, discussions with the author, November 2008.

p. 60: "fishing and shooting seals": Douglas, "Sketch" (1914), 59.

p. 60: "ascent was easier": ibid.

p. 62: "unconscious of time": ibid., 60.

p. 62: "[t]his plant might be named": Scouler, "Journal,", 197.

p. 63: "I have as far as in my power": Douglas to Hooker, October 3, 1825, Hooker Papers.

p. 63: "In my walks in the Forest": ibid.

p. 64: "Gladly would I spend a few days": Douglas to Clinton, October 3, 1825, Clinton Papers.

p. 66: Routes between Willapa Bay and Grays Harbor: Ray Gardner, discussions with the author, November 8–9, 2008.

p. 70: "aided by a head and horns": Douglas, "Observations . . . North American Mammalia," 330–32.

p. 71: "Unquestionably this is the most splendid specimen": Douglas to Scouler, April 3, 1826, Hooker Papers.

Chapter 5
Unless otherwise noted, all quotes are from David Douglas, *Journal* (1914), 157–78.

p. 73: hoping "it may not incur his displeasure": Douglas to Hooker, March 24, 1826, Hooker Papers.

p. 74: "an enormous indulgence": Douglas, "Sketch" (1914), 62.

p. 74: "I adapt my costume": Douglas, "Sketch," CBM, 106.

p. 76: Douglas to Hooker, March 24, 1826: Hooker Papers.

p. 76: Samuel Black's checkered career: DCB; Simpson, Character Book, 192–93.

p. 77: Douglas to Scouler: *Edinburgh Journal of Science* 6 (January 1827): 116–17.

p. 79: "with infinite gratification do I mention": Douglas to Simpson, April 14, 1826, HBCA, D.4/119, fo. 36.

p. 79: "Not knowing what to gather first": Douglas, "Sketch," CBM, 106.

p. 79: *Lomatiums* as tribal food source: Hunn, *Nch'I=Wana*, 99–115, 170–76.

p. 79: John Work bio: Work, "Journal," 26–29.

p. 80: Thompson Rapids: Nisbet, *Sources of the River*, 189–90.

p. 81: "the whole stream is precipitated": Douglas, "Sketch," CBM, 109.

p. 83: Spokane House: Nisbet, *Sources of the River*, 159.

p. 83: Mixed-blood culture: Nisbet, *Visible Bones*, 204–22.

p. 83: Finlay family lore: Chalk Courchane, discussions with the author, 1995–2005.

p. 85: "Their voice is the same as the sheep": Douglas, "Observations . . . North American Mammalia," 330–32.

p. 88: Bluebell Mine: Howay, *British Columbia*, 468; Phillips-Wolley, "Mining Development," 300.

p. 88: Anderson to Simpson: April 17, 1850, HBCA D.5/28.

p. 88: "summary of the year's events": HBCA Piece 32, 1847; HBCA Piece 34, 1848.

p. 88: "Servants Deceased": HBCA Piece 38, 1850.

p. 88: Marie Josephte "Josette" Finlay: Courchane, personal communication, December 9, 2007.

Chapter 6
Unless otherwise noted, all quotes are from David Douglas, *Journal* (1914), 179–212.

p. 91: "As soon as our boats": Douglas, "Sketch," CBM, 112.

p. 92: "No language can convey": ibid.

p. 93: "two species of ants": probably carpenter and harvester ants. Dr. Laurel P. Hansen, discussions with the author, May and September 2008.

p. 94: "I am under the necessity": Douglas, "Sketch," CBM, 114.

p. 94: "that I should write to England in the morning": ibid.

p. 95: "Never in my life did I feel": ibid.

p. 95: "June 15th, Thursday. My eyes": ibid., 115.

p. 95: "somewhat rancid": ibid.

p. 96: "strange species of rat": ibid.

p. 97: "This article, being, as it were": ibid., 114.

p. 98: The Nose: Hunn and Morning Owl, *In a Land*, 228.

p. 96: "the snows of which": Douglas, "Sketch," CBM, 115.

p. 101: "it is not to be wondered": ibid., 117.

p. 102: "I will engage he does not forget": ibid., 118.

p. 103: "the burning deserts of Arabia": ibid., 119.

p. 104: "The spot . . . where Lewis and Clarke built their canoes": Moulton, *Journals*, 4: 24.

p. 104: "my companion and friend": Douglas, "Sketch," CBM, 119.

p. 104: "a remarkable spring": ibid., 119–20.

p. 105: Lake Waha hydrology: Russell, "Geology and Water Resources," 79–81.

p. 105: "the poor man of language": Douglas, "Sketch," CBM, 120.

p. 105: John Work's account of incident: Work, "Journal," 34.

p. 108: Indian hemp plants: John Ross, discussions with the author, December 2008.

p. 108: Initials in grand fir trunk: Geyer, "Botanical Info," 296.

p. 109: Cedars for building canoes: Nisbet, *Sources of the River*, 183–84. This stand of cedars was on Mill Creek, just north of the present town of Colville, Washington.

Chapter 7
Unless otherwise noted, all quotes are from David Douglas, *Journal* (1914), 212–42.

p. 117: Two twenty-foot logs: Mayne, *Four Years*, 409.

p. 118: "requisite for what I call My Business": Douglas, "Sketch," *CBM*, 125.

p. 118: "highly unfavourable to botanizing": Douglas, "Sketch," *CBM*, 125.

p. 118: Setting fires to the ground: Boyd, *Indians, Fire, and the Land*, 94–138.

p. 119: "Pasture is rarely found": McLeod, Journal, October 2, 1826.

p. 119: "Thus we live literally hand to mouth": Douglas, "Sketch," *CBM*, 126.

p. 119: Paper wasp and yellow-jacket nests, filled with nutritious larvae: Dr. Richard Zack, discussion with the author, November 2008.

p. 120: birth of baby girl: McLeod, Journal, October 2, 1826.

p. 120: Sugar pine range: Arno, *Northwest Trees*, 19–20.

p. 121: Grizzly bear skin "to use as an under robe": Douglas, "Sketch," *CBM*, 126.

p. 121: "fortunately his clothing was so old": ibid.

p. 122: Trail between Willamette and Umpqua: Sweetser, "David Douglas" and "Members of Plant Kingdom."

p. 124: "always keep along the skirts of the sea": Douglas, "Sketch," *CBM*, 127.

p. 124: "The baggage which mine carried": ibid., 127–28.

p. 126: "but in the eagerness of pursuit": ibid., 128.

p. 127: "strong enough to secure the largest Buffalo": ibid., 129.

p. 128: "where the kind inhabitants kindled my fire": Douglas, "Sketch" (1914), 68.

p. 128: "to add to my miseries, the tent was blown down": Douglas, "Sketch," *CBM*, 129.

p. 129: Damaging effects of mercurous chloride in calomel: Janet Brown, *Charles Darwin*, I: 279, 488.

p. 129: "being quite satisfied that this conduct was prompted by fear": ibid.

p. 130: "endeavored to knock off the cones": ibid.

p. 130: "As much as possible I endeavored to preserve my coolness": ibid.

p. 131: "made the quickest possible retreat": ibid.

p. 131: "the tree which nearly cost me so dear": ibid.

p. 131: "a substance which, I am almost afraid to say": ibid., 103.

p. 133: "It is a relief": ibid., 132.

p. 133: "We have been entertaining one another": ibid., 133.

p. 138: George Barnston: DCB; Simpson, Character Book, 230–31.

p. 138: "the man of science": Barnston, "Abridged Sketch," 207.

p. 138: "brought out in full glow": ibid.

p. 138: "he was a perfect enthusiast" and condor behavior: ibid., 208.

p. 140: "They build their nests": Douglas, "Observations on the *Vultur*," 328–330.

p. 141: Meriwether Lewis on sewelel: Moulton, *Journals* 6: 351–54.

p. 142: Emma Luscier on sewelel: Harrington, Papers, mf 17.0398.

Chapter 8
Unless otherwise noted, all quotes are from David Douglas, *Journal* (1914), 242–94.

p. 143: "Though I hailed the prospect": Douglas, "Sketch," CBM, 134.

p. 144: Ermatinger version of lost gun: Ermatinger, "York Factory Express," 72.

p. 148: William Clark at upper reaches of Clark Fork–Pend Oreille system: Moulton, *Journals* 6: 216, 219.

p. 148: Ermatinger bartered for pas d'ours: Ermatinger, "York Factory Express," 77.

p. 148: Sturgeon-nosed canoes of the Lakes tribe: Bouchard and Kennedy, "Report," 55–58.

p. 149: "Narrows of Death": Douglas, "Sketch," CBM, 135.

p. 150: "On beholding the Grand Dividing Ridge": ibid.

p. 151: "probably the first ever eaten at this place": Ermatinger, "York Factory Express," 78.

p. 153: "The difference of climate and soil": Douglas, "Sketch," CBM, 137.

p. 157: "Poor McDonald was thus situated": ibid., 138.

p. 157: "bound up his wounds": ibid.

p. 156–57: Finan McDonald's buffalo encounter: Ermatinger, "York Factory Express," 87–89.

p. 158: Richardson's report of McDonald's death: Richardson, *Fauna*, 281.

p. 161: Spokane Garry at Red River: Drury, "Oregon Indians."

p. 164–65: Hudson Bay ordeal: Drummond, "Sketch of a Journey," 216–18.

p. 165: Douglas was confined to his bed: Douglas, "Sketch," CBM, 140.

Chapter 9

Unless otherwise noted, all quotes are from Hooker, "Memoir," *Companion to the Botanical Magazine*, 140–46.

p. 167: Plants sent and brought back to England: Douglas, *Journal*, 326–36; Harvey, *Douglas of the Fir*, 254–60; and Mitchell, *David Douglas*, 210–22.

p. 167: "His appearance one morning": Booth, "Memorandum."

p. 168: "Understanding upon my return": Douglas, "Account of . . . *Pinus*," 497.

p. 168: Douglas's account of sugar pine: ibid., 497–500.

p. 168: "a crimson substance": ibid., 499.

p. 169: Distribution of sugar pine seeds: Horticultural Society Council Minutes, March 1, 1828.

p. 170: "His company was now courted": Booth, Memorandum.

p. 170: tales of running from hostile Indians: Bentham, *Autobiography*, 316–17.

p. 170: "Flattered by their attention": Booth, Memorandum.

p. 170–71: Mariposa lily paper: Douglas, "Account of . . . *Calochortus*," 275–80.

p. 170: Lindley on mariposa lily: Lindley, *Edwards's Botanical Register* 14: 1152.

p. 171: "every assistance and facility": Horticultural Society Council Minutes, March 1, 1828.

p. 171–72: "This was done from the best of motive": Booth, Memorandum.

p. 172: "has much in his head": Hooker to Richardson, September 13, 1828, Hooker Papers.

p. 173: *Some American Pines*: Douglas, *Journal*, 338–49.

p. 173: Linnean Society nomination: Library, Linnean Society, London.

p. 174: "He is quite a *sauvage*": Bentham, *Autobiography*, 317.

p. 174: "quite disgusted with the manner in which he spoke": memo, Lindley Library, Royal Horticultural Society, London.

p. 174: "expecting they would be a treat for Mr. Sabine": Booth, Memorandum.

p. 175: "Such a meeting": Douglas to Hooker, June 17, 1828, Hooker Papers.

p. 176: "I am indebted for the accompanying figure": Scouler, "Remarks," 308.

p. 176: Condor life history: Douglas, "Observations on the *Vultur*," 328–30.

p. 176: Account of deer and sheep: Douglas, "Observations on Two Undescribed Species," 331–32.

p. 176: "my tribute of praise to Mr. Douglas": Richardson, *Fauna*, 334.

p. 177: "This beautiful and highly interesting species": Bell, "Description," 106.

p. 177: "modern herpetologists have found": William Leonard, personal communication, October 10, 2008.

p. 177: "ordered that all Fees . . . be remitted": Minutes, Linnean Society council, June 24, 1828.

p. 178: Douglas assists Sabine with dip needle: Sabine, "Report," 82.

p. 178: "He has brought a complete sample set": Hooker to Richardson, September 13, 1828, Hooker Papers.

p. 179: Daniel Macnee portrait of Douglas: Library, Linnean Society, London.

p. 179: "a fisherman was on the coach": Douglas to Hooker, November 18, 1828. Hooker Papers.

p. 179: "Faults Mr. Douglas may have had": *Cottage Gardener*, 263.

p. 180: Atkinson sketch: Library, Linnean Society, London.

p. 180: Colonial Office and boundary line: Douglas to R. W. Hay, Colonial Office Papers, class 6, volume 6, fos. 807–13. Public Record Office, London.

p. 181: Meeting Francis Boott: Douglas to Hooker, November 18, 1828. Hooker Papers.

p. 181: A paper on grouse and quail: Douglas, "Observations of . . . *Tetrao and Ortyx*," 133–49.

p. 182: "I look forward at no distant period": ibid., 149.

p. 182: "the London climate kills me": Douglas to Hooker, November 18, 1828, Hooker Papers.

p. 182: "The beginning was bad": Douglas to Hooker, December 1, 1828, Hooker Papers.

p. 182: "Soured in his temper": Booth, Memorandum.

p. 183: "As it is doubtful": Douglas to Vigors, December 29, 1828, Douglas Papers. Royal Horticultural Society, London.

p. 183: Currant paper: Douglas, "Account of . . . *Ribes*," 509–18.

p. 184: "of such importance do we consider it": Lindley, *Edwards's Botanical Register*, 16: 1349.

p. 184: "I am doing my best to get finished": Douglas to Hooker, June 7, 1829. Hooker Papers.

p. 184: "shrieks were dreadful": Elliott, *Royal Horticultural Society*, 114.

p. 185: R. W. Hay of Colonial Office and expenses: Harvey, *Douglas of the Fir*, 163.

p. 186: "conquering every difficulty": Abbott, "Observations," 472.

p. 187: "we may now anticipate": *Zoological Journal* 4: 355–56

p. 187: "The route of Franklin, Richardson, and Drummond": Douglas to Hooker, September 14, 1829. Hooker Papers.

p. 188: Mistake on the elevation of Mount Brown: Harvey, *Douglas of the Fir,* 130–44.

p. 188: "This may be true": Douglas to Hooker, September 14, 1829. Hooker Papers.

p. 189: "Douglas made himself some enemies in the Columbia": Richardson to Hooker, January 28, 1830. Hooker Papers.

p. 189: "I cannot tell you how pleased I am": Douglas to Hooker, October 27, 1829. Hooker Papers.

p. 190: "I hope, ere the whole of the Flora is printed": ibid.

p. 190: "I have applied to Mr. Sabine for a copy": ibid.

Chapter 10
Unless otherwise noted, all quotes are from Hooker, "Memoir," *Companion to the Botanical Magazine,* 142–62.

p. 191: "During the whole passage": Harvey, *Douglas of the Fir,* 165.

p. 191: "How beneficial it is for a person like me": Sabine, Observation, 8.

p. 193: "Some of my fashionable London friends": Douglas to Hooker, November 20, 1831. Hooker Papers.

p. 193: "We were glad to see again": Barnston, "Abridged Sketch," 267.

p. 193: Fort Vancouver, 1830: McLoughlin, *Letters,* ed. by Rich, 92.

p. 194: Alexander McKenzie murder: ibid., 57, 63–64.

p. 194: survey equipment: Barnston, "Abridged Sketch," 268.

p. 194: "during his leisure hours": ibid.

p. 194: "He would frequently spring up": ibid.

p. 195: Blue Mountain journey: ibid., 268–69.

p. 196: Lizard collecting: ibid., 269.

p. 196: "I wish you had been with me": ibid., 271.

p. 196: visit with Chumtalia: ibid., 270.

p. 197: Alexander McLeod: DCB; McLoughlin, *Letters,* ed. by Barker, 76.

p. 197: William Johnson: Reed, "William Johnson."

p. 197: "The day however was not so good": Barnston, "Abridged Sketch," 271.

p. 198: Santiam River losses: ibid., 272.

p. 198: "[p]oor Chumtalia is since dead": ibid.

p. 199: Intermittent fever: McLoughlin, *Letters,* ed. by Rich, 166.

p. 199: "Douglas was ill like others": Barnston, "Abridged Sketch," 270.

p. 200: Intermittent fever as malaria: Boyd, *Coming of the Spirit,* 84–115.

p. 200: George Barnston wondered: Barnston, "Abridged Sketch," 270.

p. 201: A load of sawn timber and salted salmon: McLoughlin, *Letters,* ed. by Barker, 159–62.

p. 201: Stock of personal items: Fort Vancouver Accounts, HBCA, B.223/c/1, fos. 420–23.

p. 201: "To the Commander of any": McLoughlin, *Letters,* ed. by Barker, 171.

p. 202: Beechey to Hartnell: Dakin, *Lives of William Hartnell*, 169–70.

p. 203: Pinus sabiniana paper: Douglas, "Account of . . . *Pinus*," 747–48.

p. 203: "well meant on my part": Douglas to Hooker, October 23, 1832. Hooker Papers.

p. 204: "Dr. Douglas set my arm": Davis, *Seventy-five Years*, 206–7.

p. 204: "Their localities are determined": Barnston, "Abridged Sketch," 277.

p. 204: "I lived almost exclusively with the fathers": ibid., 274–75.

p. 207: Company of Foreigners: Dakin, *Lives of William Hartnell*, 202–6.

p. 208: "The ladies are handsome": Barnston, "Abridged Sketch," 275.

p. 208: "*Calida Fornax*": at that time some travel writers used the Latin term *Calida fornax*, "hot furnace," to describe the weather and supply a name for Spanish California.

p. 208: "In no place on the globe": Barnston, "Abridged Sketch," 276.

p. 209: Douglas's resignation: Horticultural Society Council Minutes, 1833.

p. 210: precaution "surely in these times is necessary": Douglas to Hooker, October 23, 1832. Hooker Papers.

Chapter 11

Unless otherwise noted, all quotes are from Hooker, "Memoir," *Companion to the Botanical Magazine*, 142–62.

p. 211: "Who was on board also but our old friend David": McDonald to Ermatinger, February 20, 1833, Archives of Canada, Ottawa.

p. 213: "I carry your letter about in my note-book": Douglas to Hooker, October 23, 1832. Hooker Papers.

p. 213: Letter to young William: ibid.

p. 213: specimens for John Scouler: ibid.

p. 214: "eating and drinking the good things": Wyeth, "Correspondence," 178.

p. 214: Greenwich Boys: McLoughlin, *Letters*, ed. by Barker, 260.

p. 214: George Roberts's description of Douglas: Roberts, "Letters," 188–89.

p. 215: "Captain Sabine goes so far as to say": Douglas to Hooker, October 24, 1832. Hooker Papers.

p. 216: "In the pursuit of any subject": Barnston, "Abridged Sketch," 277.

p. 216: "a sturdy little Scot": MacDonald, *Ranald MacDonald*, 136–37.

p. 216: Supplies from Fort Vancouver: Fort Vancouver Accounts, HBCA, B.223/c/1, fos. 420–23.

p. 217–23: Fur trade brigade trail between Fort Okanagan and Fort St. James: William C. Brown, *Old Fort Okanagan*; Creech, *Brigade Trails*; and Gibson, *Lifeline*, 41–98.

p. 217–18: Cattle in the New Caledonia District: Gibson, *Lifeline*, 187; McLoughlin, *Letters*, ed. by Barker, 241–42.

p. 218: A square, leather-bound journal: Douglas, Field Sketches.

p. 220: "On a visit of David Douglas to Saml Blacks post": Roberts, "Letters," 191.

p. 222: Eighteen horses drowned: Gibson, *Lifeline*, 85.

p. 224: "the situation of the post": Gibson, *Lifeline*, 64.

p. 226: "met with another of those unfortunate accidents": Barnston, "Abridged Sketch," 278.

p. 227: Meredith Gairdner: Harvey, "Meredith Gairdner."

p. 227: "follow the plan of Douglas": Gairdner to Hooker, August 31, 1833, Hooker Papers.

Chapter 12
Unless otherwise noted, all quotes are from Hooker, "Memoir," *Companion to the Botanical Magazine*, 161–82.

p. 229: "Whether I return through the Russian Empire": Barnston, "Abridged Sketch," 277.

p. 229: Dryad "nearly a wreck": Dakin, *Lives of William Hartnell*, 172–74.

p. 229: Travels around San Francisco Bay: ibid.

p. 230: Coordinates to Governor Figueroa: Harvey, *Douglas of the Fir*, 213.

p. 230: Plate of volcanic plumbing: Lyell, *Principles of Geology*.

p. 239: Letter to Mrs. Richard Charlton: Douglas, Letter, 101–3.

p. 240: "the Land o' cakes": Dakin, *Lives of William Hartnell*, 173.

p. 247: Meredith Gairdner re: doubts about Douglas's accident: Gairdner to Hooker, November 19, 1835. Hooker Papers.

p. 247: Charles Wilkes visit to islands in 1841: Wilkes, *Narrative*, 204–5.

p. 248: Archibald McDonald on Douglas's death: McDonald to Hooker, April 15, 1836, Hooker Papers.

p. 248–49: John Douglas on Douglas's death: John Douglas to Hooker, March 23, 1835, Hooker Papers.

p. 249: William Booth on Douglas's death: Booth, "Some Account," 272.

p. 250: Ranald MacDonald on Douglas: MacDonald, *Ranald MacDonald*, 136–37.

p. 250: George Barnston on Douglas: Barnston, "Abridged Sketch," 121.

Epilogue
p. 252: Etching of blazing star: Hooker, *Flora*, 3: plate LXIX.

p. 252: Contributions to silviculture: Mitchell, *David Douglas*, 185–201.

p. 254: Joseph Hooker and Darwin: Janet Brown, *Charles Darwin*, 1: 418, 452.

Index

A

Aberdeen, Washington, 67
Albany, New York, 14, 18
Albany, Oregon, 120
Albatross, 30, 33, 54
Alcoholic beverages, 55, 65, 133
    beer, 162
    rum, 63, 65
    spruce beer, 32
    wine, 27, 201, 208, 216
Alder, 43
Aleutian Islands, Alaska, 187, 206
American presence in Northwest, 43, 45, 193, 200
Amherstburg, Ontario, 15, 17
Anderson, Alexander, 88
Anemone, 98, 153
Ann Marie (ship), 2, 9
Ants, 93
Antelope bitterbrush, 51, 76, 80, 93, 95
Aplondontia. See Mountain Beaver
Apple, 9, 11, 24
    cider, 21
Arbutus, trailing, 23
Arden, Washington, 82
Arkansas River, 13, 19
Arrow Lakes, 148–49
Artemesia, 68
Asbestos thread, 194
    See also Surveying Instruments
Aspen, 222, 223
Asphodel, 46
Assiniboine River, 160
Astoria, Oregon, 34
Astronomy, practical, 138, 185, 191, 196, 215, 225, 229
Astronomical observations, 186, 192, 194, 195, 196, 201, 215, 216, 225, 229, 235, 249
Athabasca Pass, ix, xi, xiii, 151, 188

Athabasca River, 152, 153
Atkinson, William, 9, 175
Atlantic Ocean, 3, 9, 21, 27–30, 191
Ava fish, 234
Ax, 61, 103, 126, 132, 154

B

Baby blue-eyes, 203
Back, George, 163–64
Badger, 78, 106
Baker Bay, 34, 37, 38, 55, 139
Bald Mountain, 99
Banana, 27, 231, 234
Banks, Sir Joseph, 7
Barley, 193
Barnston, George, 138, 139–40, 143, 193–96, 199, 200, 208, 226, 233, 250
Barometer, 187, 233, 238
    See also Surveying instruments
Bartram, John, 19
Bartram, William, 19–20, 172, 173
Basalt, 80, 98, 103, 105, 107, 253
Basketry, 56, 62, 107, 137
Bateaux. *See* Canoe
Beads, 38, 62, 125, 137, 217
Bear-grass, 56, 61–62, 67, 109
Bears, 78, 122, 149, 204, 211
    grizzly, 85, 120, 121, 132
    pelt, 120, 121, 123, 134
Beattie, William, 4, 5, 249
Beaver, 44, 56, 57
Beech, 15, 17
Beechey, F. W., 201
Bell, Thomas, 176–77, 196
Belle Vue Point, 43
Ben Lomond (Highlands), 47, 179, 241
Bentgrass, 46
Bentham, George, 173–74, 175, 209, 246
Bentham, Jeremy, 174

Berries, xi, 32, 55, 62, 63, 66, 67, 85, 93,
104, 113, 114
dried, 38, 139
Bible, 185, 236, 240
Big-cone pine. *See* Pines
Big Bend of Columbia, 78, 217
Big Stone. *See* Grosse Roche
Bighorn sheep, 13, 80, 85, 86, 113, 123,
152, 153, 156, 176
Billy (terrier), 96, 197, 204, 208, 217,
225–26, 229, 234, 241–43, 245, 246,
248
Birch, 50, 97, 147, 149
Biscuits, 54, 60, 97, 216
Biscuitroot, 48, 79
See also Lomatium
Bison. *See* Buffalo
Bitterroot, 83, 254
Black, Samuel, 76–77, 93, 95–97, 101,
113, 220–21, 226
Blackberries, 122
Blackfeet Indians, 88–89
Blalock Island, 144
Blalock Mountain, 253
Blanket, x, 49, 54, 60, 66, 67, 74, 82, 86,
93, 96, 100, 112, 114, 118, 122, 124,
128, 131
trade item, 62, 121
Blazing-star, 52, 203, 251–52
Blood-letting (phlebotomy), 126, 157
Blue Mountains, 77
Douglas's travels in, 96–100, 195, 226,
253
map of, 72
Bluebell Mine, 88
Boat Encampment, 150–51, 180
Bobcat, 71
Boki, Liliha (Hawaiian governess), 192
Bonaparte River, 222
Bonneville Dam, 48
See also Cascades of the Columbia
Booby, blue-footed, 33
Books, 11, 19, 20, 31, 48, 160, 172, 178, 192
read by Douglas, 5, 16, 26, 178, 187,
189, 230

Booth, William Beattie, 5, 9, 26, 167,
169–70, 171–72, 174–75, 182, 249
correspondence with Douglas, 32, 63,
94
Boott, Francis, 181
Botanical collectors, 7–8, 153, 252, 255
Botanical Magazine, 249
Botany, 8
See also Taxonomy
Bow and arrow, 46, 70, 75, 105, 110, 129,
130
Boundary, United States and Great
Britain, 43, 180
Brackenridge, J. D., 247–48
Brazil, 28–29
Breadfruit, 231, 234
Briscoe, Henry, 15, 16, 17
British Colonial Office, 180, 185, 249
Brodeia, 87
Bromeliad, 2, 231
Broomrape, 51
Brown, Robert, 8, 26, 99, 173, 175,
178
Brownian motion, 178
Brown's peony. *See* Peony, Brown's
Buckwheat, 93, 174
Buffalo, 127, 156–58, 162
meat, 80, 82, 87, 93
robe, 124
Buffalo berry, 81
Buffalo, New York, 14, 17
Bunchgrass, 83, 95, 106, 220, 253
Burr, Aaron, 10
Buttercup, 79
Butterflies, 143
Buttons, 139, 217

C

Cacao, 8
Cactus, 95, 144
California, 141, 169, 200, 201
Douglas's travels in, 201–8, 211
California laurel, 123–24, 125, 133
Calipio (Hawaiian guide), 236–39, 240
Calochortus. *See* Lily, mariposa
Calomel, 128–29

Camas, 45, 137, 144, 254
  as food, 45, 83, 125, 134
  ovens, 45
Campion, 159
Cancer-root, 15, 17
Canoes and Douglas, 42–43, 46, 49, 52,
  57, 59, 65, 67, 74–78, 80, 86, 91–93,
  102, 103, 112–14, 125, 128, 137–39,
  143–45, 147–50, 153–54, 156, 159–60,
  162, 163, 194–95, 198, 217, 225–26
  bateau, 49–50, 92, 118, 223
  birchbark, 15, 153, 223, 224
  construction of, 46, 104, 109, 149, 150,
    236
  fur trade, 42, 44, 47, 64, 68, 74, 91–93,
    143
  dugout, 37, 55, 65–66, 126
  sturgeon-nosed, 148, 149
Canoe River, 150
Cape Cod, 10
Cape Disappointment, 34, 38, 54, 141,
  193, 201, 229
Cape Horn, 30, 64
Cape Verde Islands, 28
Caribou, 148, 154, 255
Cardinal, Jacques, 152–53
Carlton House, 158
Carr, Nancy, 19, 20
Cascades of the Columbia, 48, 50, 58,
  59–60, 69, 74, 103, 114, 143, 195
Cascade Range, 57, 216, 219, 227
Castlegar, British Columbia, 148
Cattail, 51, 56
Cattle
  California, 204
  Columbia River, 44, 74, 80
  drive to Fraser River, 217–24
  Hawaiian, 231, 242, 243, 244, 246,
    248
  Hawaiian hunters, 231, 242, 244, 248
  Hawaiian traps, 242–43, 244, 245,
    246–47, 248
Cayuse Indians, 48, 75–76, 95, 100
  guide, 96, 97–101
Cedar, 41, 50, 87, 104, 108, 149
  bark, 50, 56, 68, 125, 149

Celilo Falls, 51–52, 75, 113, 143
Centrenose (Umpqua chief), 125, 127
  son of (guide), 127–32
Champoeg, Oregon, 118, 137
Challenger (ship), 245, 246
Chamokane Creek, 110
Charlton, Richard, 208–9, 230, 240, 244,
  245–46
Charlton, Mrs. Richard, 208–9, 230,
  239–40, 247
Cheney, Washington, 107
Chehalis River, 67
Chehalis Indians, 67
Cherry Creek, 221
Chevreuil, 70
Chickadee, 147
Chicken, 234
Chickweed, 159
Children, 7, 64, 159, 160, 161, 236, 241
  of fur traders, 44, 45, 83, 102, 118, 120,
    155, 216, 250
  tribal, 40, 56, 62, 125, 161, 195–96,
    254
Chile, 9, 29, 31
Chinquapin, 121
Chinook jargon, 39–40, 58, 70, 75, 113,
  125
Chinook Indians, 37–38, 39, 40, 46–47,
  51, 55–56, 64–67, 176, 188
  See also Basketry, Cradle board,
    Weaving, Women
Chiswick Gardens, 8, 9, 26, 104, 170,
  171, 174, 184, 200, 252
  Douglas and, 23, 46, 53, 93, 167–68,
    169, 172, 173, 174, 180, 183
Chocolate, 65
Chokecherry, 149
Chronometer, 187, 245, 246, 247
Chumtalia (Cascades chief), 59–61, 114,
  195, 198–99
  brother of, 59–60, 114
Church services, 29, 161, 162, 236
Clallam Indians, 194, 197
Clark Fork River, 88
Clark, William (Corps of Discovery), 52,
  148

Clark, William (Juan Fernandez Island), 31–32
Clarkia, 52, 184
Clarkston, Washington, 103
Clatsop Indians, 55
Clearwater River, 103, 105, 180
Clematis, 189
Cleveland, Oregon, 128
Clinton, DeWitt, 14, 17, 18, 21, 24, 64
Clover, 49
Cloak, 67, 74
    Hawaiian feather, 236
Clothing, 121, 157
    Chinook, 39, 141, 176
    Douglas's, ix, 6, 48, 74, 109, 110, 112, 113, 115, 118, 122, 124, 129, 149, 150, 151, 154, 159, 160, 163, 201, 216, 243
    Umpqua, 125
Coa bark, 233
Cock de Lard, 104–5
Cockqua (Chinook chief), 55–56, 62, 139, 142, 146
Cockroach, 209, 213
Coffee, 216, 231
Columbia District, xi, 42, 45, 73, 86, 193
Columbia Gorge, 51, 61, 74, 114
Columbia Plateau, 51, 70, 73, 251
Columbia River, ix, xi, 26, 180
    bar, 34, 188, 193, 201, 211
    Douglas's travels on lower river, below the Dalles, 35–51, 53–54, 59–65, 113–15, 138–39, 141–42, 193, 196–97, 201, 211–14
    Douglas's travels on middle river, between the Dalles and Kettle Falls, 51–52, 75–81, 86–87, 91–93, 102–3, 110–13, 143–45, 194–95, 217, 226
    Douglas's travels on upper river, above Kettle Falls, 87, 147–51
Colville River, 108
Colville Valley, 82–83, 87–89, 109
Comcomly (Chinook chief), 39, 44, 65, 214
Committee's Punch Bowl, 152
Compass, 178, 187, 192, 245

Condors, California, 70, 120, 139–41, 176
Connolly, William, 93, 97, 102
Continental Divide, ix, x, xii, 150–52, 154, 188
Copper, 38, 118
Corn, 8, 15, 134, 193
Cornwall, England, 21
Cottonwood, 106
Cougar, 213
Coulter, Thomas, 206, 207
Cow Creek, 106
Cow parsnip, 70, 105
Cowlitz Indians, 67
Cowlitz River, 47, 67, 68, 215–16
Coyote, 146, 156
Coyote (folklore), 146
Cradle board, Chinook, 176
Craig Mountain, 104
Cree Indians, 45, 83
Crow, 69
Crowberries, xi
Cruz Bay, Juan Fernandez Island, 32
Cumberland House, 158–59
Curlew, 81, 87, 145
Currant, 40, 67, 75, 84–85, 98, 104, 105, 108, 183–84, 203
Cypress, 19

D
Dak-elh Carrier Indians, 224
Dalles, 51–52, 54, 75, 102, 113–14, 143, 170, 195, 251, 252
Dalles des Morts, 149, 150
Dana, James, 247
Dancing, 153, 156, 184
    Chinook, 39, 55
    grouse, 144–45, 181
    Nez Perce and Palouse, 104
Darwin, Charles, 33, 254–55
Davidson, Captain, 115
Davis (Hawaiian rancher), 242, 246
Davis, William Heath, 204
Dease, John, 78, 80, 82, 84, 87, 109, 110, 112, 145
Dease River. *See* Kettle River
Death camas, 109

Deer, 70, 118, 125, 126, 127, 137, 176, 213
  coastal black–tailed, 57, 70, 123, 127
  mule, 70
  white-tailed, 70, 122, 127, 133
Delphinium, 207
Desert parsley. *See Lomatium*
Detroit River, 14, 15, 17
Dick, William, 13, 19
Diell, John, 241, 243–45
Dip needle, 178, 187, 192, 229, 246
Divide, Oregon, 122
Dogs, 10, 15, 71, 103, 146, 154
  and Douglas, 96, 134, 197, 200, 204, 234
  See also Billy
Dogwood, Pacific, 46
Dollond, John, and Sons, 187, 195
Dormouse, 58
Douglas, David
  apprenticeship, 5
  childhood, 3–4
  death and inquest, 243–46
  death, questions about, 246–48
  employment, 4, 5, 6, 7
  eyesight, 77, 81, 95, 101, 128, 179, 185, 194, 205, 217, 237, 238
  family, 3, 159, 179, 185
  father, 3, 159
  funeral, 245, 246
  journals, xii, 1, 16, 48, 63, 94, 151, 172, 189, 191, 218, 225, 229, 240, 249
  knee injury, 64–65, 66, 67, 69, 73
  lodging, at Fort Vancouver, 43, 48, 56, 69
  in London, 169
  marksmanship, 55, 58, 66, 77, 122, 130, 133
  medical reputation, 204
  physical description, 179–80, 214–15, 216
  portraits of, 179–80
  religion, 119, 149, 185, 232–33
  salary, 183, 185
  scientific writings, 25, 168, 170, 173, 176, 177, 181, 183, 203

sketch maps, 218–24
son of, 88–89
tin box, x, 124, 143, 150, 151, 155, 168
travels in eastern United States and Canada, 10–21
travels in west. *See* specific locales
tributes to, 249–50
voyage to New York City, 1–3; 9–10
voyages to Pacific Northwest, 26–34; 191–93
  See also Hudson's Bay Company, London Horticultural Society,
Douglas, John (brother), 3, 4, 5, 26, 63, 102, 138, 159, 185, 240, 248–49
Douglas fir, 40, 41, 61, 117, 173, 252
Dover's powder, 86
Drain, Oregon, 123
Dryad (ship), 109, 112, 115, 117, 200–1, 205, 229–30
Dryas, 152
Drummond, Thomas, 152, 155, 158, 163–65, 189
Ducks, 21, 38, 66, 135, 136
Duckweed, 111
Duke of Devonshire, 122

E
Eagle
  bald, 38, 55, 60, 69, 159, 162
  calumet. *See* golden eagle
  golden, 155–56, 159, 163
  quill, 109
Eagle Creek (Oregon), 60
Eagle (ship), 189, 191–93, 200, 211–14
Eclipse, lunar, 215
Edible plants, 40, 45, 46, 48, 54, 56, 66, 79, 83–84, 87, 89, 125, 131, 134, 137, 139, 144, 169, 171, 220, 234, 253
Edward's Botanical Register, 170, 174
Eggs (bird), 81, 82, 92, 140–41, 181
El Camino Real, 204, 207
Elderberry, 105
Elk, 48, 121–22, 127, 156
Elk Creek (Oregon), 123, 126, 134
Entiat River, 112

Epiphyte, 231
Equator, 28
Erie Canal, 14, 17
Ermatinger, Edward, x, xi, 142, 143–53,
 156–58
Eugene, Oregon, 121
Evening primrose, 49, 57, 68, 77
Eyeglasses, ix, 47

F

False asphodel, 67
Fannaux, 134–37
Fern, 78, 128, 133, 192, 231, 233
 bracken, 120, 122
 sword, 56
Fever and ague. *See* Intermittent fever
Figueroa, Governor, 207, 230
Finlay, David, 88–89
Finlay, Jaco, 82–85, 107
 sons of, 82–86
 wife, Teshwentichina, 83
Finlay, Marie Josephte, 88
Finlayson, Duncan, 208, 210
Fir, 41, 108
 bristlecone, 207
 Douglas. *See* Douglas fir
 grand, 122, 200, 216
 noble, 61, 198, 200
 Pacific silver, 61
Fire, for land management, 83, 118–19
Firearms, 134, 144, 206, 221, 243
 Douglas and, 48, 55, 66, 75, 82–84, 96,
 110, 111, 118, 123, 129, 130, 131,
 133, 135, 187, 204
Fish, xii, 54, 163, 218, 222, 234
 See also individual species
Fishing, 132, 206
 Douglas, 4, 6, 50, 54, 60, 179, 213, 254
 Tribal, 43, 46, 50, 66, 81, 107–8, 110,
 114, 127, 132
Flathead River, 109
Fleas, 59, 93, 236
Flint and steel, xi
Flora Americae. See Pursh, Frederick
Flora Boreali-Americana. See Hooker,
 William

Flushing, New York, 11–12, 21
Flute, 149
Flying fish, 28
Foamflower, 40
Forget-me-not, 47
Fort Alexander (Lake Winnipeg), 160
Fort Alexandria (Fraser River), 223, 226
Fort Assiniboine, 153, 154
Fort Colvile, 78, 80–82, 83, 86–88, 91,
 107, 109, 145–147, 151
Fort Edmonton, 154–56
Fort Garry, 160
Fort George (lower Columbia), 34, 38–42,
 43, 44, 47, 48, 65, 138, 201, 211
Fort George (upper Fraser), 224, 225, 226
Fort George Canyon (Fraser River), 225
Fort Kamloops, 220, 226
Fort Nez Perce. *See* Fort Walla Walla
Fort Nisqually, 216, 227
Fort Okanagan, 78, 92, 107, 112, 145,
 217–18, 226
Fort Ross, 205
Fort St. James, 224–25
Fort Simpson, 224
Fort Vancouver, 43–44, 193
 as benchmark, 215
 Douglas's visits to, 43–9, 53, 56–57,
 59, 62–64, 67–71, 74, 114–15, 117,
 137–38, 139–41, 143, 193–94, 197,
 198–201, 214–16, 227
 mentioned, 52, 91, 94, 97, 102, 109, 119,
 134, 146, 175, 205, 206, 207, 220
Fort Walla Walla, 76–77, 91, 93–97, 100,
 101, 103, 113, 161, 194–96, 217, 226
Fossils, 41, 222
Fox, 70, 78, 213
Franciscan monks, 169, 204–5, 229
Franklin, John, 159–60
 Arctic expeditions, 25, 26, 68, 76, 152,
 163, 172, 187
Fraser River, 93, 154, 194, 216, 220, 223,
 225, 229
Frasera, 101, 253
Free hunters, 45
French Canadians, 45, 99, 118
 See also Voyageurs

Frittilary, 46
Frog, 106
Fruit culture, 9
Fulmar, 30
Fulton vegetable market, 11
Fur packs, 44, 64, 91, 92, 213

G

Gairdner, Meredith, 227, 247
Galapagos Islands, 32–33, 254
Galena, 88
Gambling articles, 56
Garry, Nicholas, 146, 160, 175
Geese, 135, 136, 151, 209
Gentian, xi
Geology, 19, 41, 50, 88, 111–12, 145,
    176, 208, 230–31, 239, 247
George Bank, 10
Glacier, xi, 150, 152, 220, 222
Glasgow, Scotland, 6, 27, 47, 173,
    178–79, 210, 249
Glasgow Botanic Garden, 6
Glasgow University, 6, 27
Globemallow, 92
Goat, 231
Goat Island, 17
Goggles, 217
Gold, 87–88
Goldenrod, 120
Goodrich, Joseph, 230, 233, 239, 240,
    241, 243–45
Gooseberry, 203
Gopher, pocket, 58
Graham, Maria, 29
Grand Cote, ix
Grand Coulee, 111–12
Grande Ronde River, 99, 100, 195
Granite, 14, 29, 80, 219
"Grass Man," 75, 114
Grass, timothy, 46
Grasshoppers, 118, 119
Grave, Captain, 211
Gravesend, 26
Grays Harbor, 65, 66
Greenbrier, 17
Green Lake, 222

Greenwich, 185–86, 195
"Greenwich boys," 214
Grosse Roche (Big Stone), 145
Ground squirrel, 58, 95, 102, 117, 177
Grouse, 81, 82, 92, 145, 176, 178, 181,
    214
  blue, 82, 86–87, 92, 181
  dusky. *See* Grouse, blue
  ruffed, 181
  sage, 78, 92, 144–45, 146
  sharp–tailed, 78, 87
  spruce, x, 153, 154, 181
Gull, 54, 66
Gun. *See* Firearms
Gunflint, 56, 67
Gunpowder, 113, 137, 149, 198
Gurney, Ned, 242–43, 245, 246, 247, 248

H

Hall, Charles, 244, 246
Hamilton, Alexander, 10
Hammer, rock, 139
Hanford Reach, 77
Hanwell, Henry, 27, 28, 30, 31, 32, 34,
    37, 39, 42, 62, 63, 64
Harborseals, 60
Harrington, John, 142
Hartnell, William, 201–2, 207
Harvey, A. G., 248
Hatchet. *See* Ax
Hats, 97, 217
  Chinook, 55, 56, 62
  Douglas's, 33, 115, 135, 136, 216
Hawaii (the island), 230–40, 241–45,
    246–48
  See also Hilo
Hawaiian Islands, 192, 229
Hawaiian laborers on Columbia, 45, 118,
    199
Hawaiian Spectator, 247
Hawk, 238
Hawthorn, 80
Hay, R. W., 180, 185
Hazelnut, 113, 122, 125, 147
Hellebore, false, 121
Hemlock, 41, 61, 122

Herbarium, 7, 25, 26, 158, 181, 184
Highlands (Scotland), 6, 47, 57, 179, 213
Hilo, Hawaii, 230, 233, 239, 240, 241, 242, 243–44, 246–48
Hogs, 44, 74, 80, 231, 234
Hogg, Thomas, 11, 12, 13, 20, 21
    son of, 13
Honey, 118, 119
Honeycreeper (bird), 236
Honeysuckle, 16, 24, 40, 98
Honolulu, 208–9, 230, 240–41, 244, 245–46, 247
Honorii, John, 230–33, 234–37, 238–39
Hooker, Joseph Dalton, 7, 179, 241, 254–55
Hooker, Maria, 6–7, 76, 241
Hooker, William Jackson, 6, 7, 11, 27, 32, 165, 172, 173, 178–79, 182, 183, 227, 249
    correspondence with Douglas, 47, 48, 63, 76, 79, 159, 175, 181, 184, 185, 187, 188, 190, 193, 195, 200, 203, 205, 206, 207, 209, 213, 217, 225, 227, 240–41
    Flora Boreali-Americana, 179, 183, 187, 189–90, 211, 213, 252
    mentor to Douglas, 7, 24
Hooker, William, Jr., 7, 64, 76, 213, 241
Horses, 1, 44, 51, 69, 85, 102, 127, 145, 222
    and Douglas, 16, 46, 82, 96–100, 104–5, 107–8, 109–12, 118–24, 125–26, 127–32, 134–37, 147, 152–53, 170, 195, 219–23, 226
    as food, 95, 106, 144
    pack animals, 67, 119–24, 152–53
    trading, 103–6
Horsetail, 40
Horticultural Society of London. *See* London Horticultural Society
Hosack, David, 10, 11, 12, 14, 18, 20, 21, 24
Huckleberry, 62, 113, 114, 122
Hudson Bay, 68, 74, 78, 143, 158, 159, 163–65
Hudson River, 12, 14, 20

Hudson's Bay Company, 25, 42
    Columbia District, xii, 37, 39, 43–45, 83, 180, 194, 200, 224
    Douglas, 25, 26, 42, 185
    employees x, 44–45, 76, 100, 189, 199, 208, 221
    express, 68, 74–78, 138, 143–54, 156–58, 162–63, 216–17
    fur brigade, 44, 91–94, 102–3, 137, 156, 194
    London headquarters, 42, 146, 167, 175, 188, 248
    supply ships, 25, 26, 27, 49, 94, 185, 189, 193
Hyacinth, 87, 254
Hydrometer, 187
Hygrometer, 186

I
Ice Age flood channels, 106, 111
Iguana, 32, 33
Illum-Spokanee, Spokane chief, 85, 161
Indian hemp, 50, 107, 109, 159
Indian potato. *See* Spring beauty
Insects, xii, 27, 29, 48, 119
Intermittent fever, 199–200, 214, 227, 247
Iris, 58, 127
Iroquois Indians, 45, 63, 120
Isabella (ship), 193

J
Jackrabbit, 106
James Island, Galapagos, 32
Jasper House, 153
Jay, gray, 152
    Steller's, 69, 151
John (Diell's manservant), 241–42, 244
John Day River, 102
Johnson, William, 197–98, 201, 204, 208, 211, 215, 217–18, 225–26
Journals. *See* Douglas, journals
Juan Fernandez Island, 29, 31

K
Kalapuya Indians, 57–59, 70, 118–19, 136–37, 198

Kamloops Lake, 221
Kapapala, Hawaii, 236
Kendall, E. N., 163–65
Kennedy, John, 120, 121, 126, 134–37
Kettle, 48, 98, 112, 118, 133, 137, 144
Kettle Falls, 80, 81, 91, 147
Kettle River, 86
Kilauea crater, Hawaii, 234–35, 239, 255
Kinnikinick, 48, 66
Kittson, William, 86, 91, 92
Klamath region, 120
Knight, Thomas, 24
Knife, 56, 97, 130, 199
    Douglas's, 102, 103, 108, 111
Kohala Point, Hawaii, 241
Kootenay Lake, 88
Kootenay River (Canada), 148
Kootenai Indians, 109–10

L

La Compania Extranjera, 207
Labrador tea, 163
Lac de Chevaux, 222
Ladyslipper, *See* Orchids
Laframboise, Michel, 132–33
Lake Erie, 14, 17
Lake St. Clair, 16
Lake Waha, 105, 255
Lake Winnipeg, 159
Lakes Indians, 148, 149
Lama (ship), 210, 211–12
Lambert, Aylmer, 26, 169, 173, 176
Lamb's quarters, 46
Lance-leafed plantain, 46
Lancet, 157
Larch, Western, 80
Latitude and longitude, recorded by
    Douglas, 3, 27, 28, 186, 193, 219, 222,
    223, 230
Laudanum, 157, 239
Laurel. *See* California laurel
Lava, 232, 234, 235, 238, 239
    tubes, 236
Ledbetter Point, 66
Leggings. *See* Clothing
L'Etang, Pierre, 91–92

Lewis, Meriwether, 45, 54, 141
Lewis and Clark Expedition, xii, 13, 26,
    48, 73, 95, 104, 170, 180, 181
Lewiston, Idaho, 103
Lichen, 78, 83, 148, 233
    cakes, 83–84, 88
Lily, 62, 87, 151
    fritillary, 46
    glacier, 81, 146
    mariposa, 52, 107, 170–71, 203, 207,
        253
    See also Yellow bells
Limestone, 50, 80
Lindley, John, 26, 117, 170, 172, 173,
    174, 182, 184, 203
Linnean Society, 168, 169, 173, 175, 176,
    181, 203
Linneas, Carl, 7
Little Pend Oreille River, 82, 86, 108
Little Slave Lake, 154
Little Soap Lake, 219
Little Spokane River, 83, 107
Little Wasp (interpreter), 97–101
Little Wolf (guide), 109–10
Liverpool, 1–2
Liverpool Botanic Garden, 2
Liverpool Mercury, 248
Lizard, 195–96
    sagebrush, 196
    short–horned, 176–77, 196
    western fence, 196
Lockport, New York, 17
Locoweed, 106, 174
Locust
Loddiges, George, 173
Lomatium, 79, 87, 254
London Horticultural Society, 8, 9, 11,
    25, 26, 99, 174, 175, 189, 209–10, 249
    and Douglas, x, 7, 9, 13, 25, 42, 63,
        73–74, 167, 170, 171, 178, 183,
        185, 213, 246
    Douglas's resignation from, 209–210
    plants collected by Douglas, 24, 117,
        169, 170, 182, 183, 184
    Transactions, 23, 170
London Geological Society, 176

London Zoological Society, 176, 183, 187, 209
Long Island, 10
Long Mountain. *See* Mauna Loa
Longitude. *See* Latitude
Lotus, American, 13
Lupine, 46, 52, 56, 66, 77, 93, 98, 99, 173, 174, 253
Luscier, Emma, 142
Lutke, Fyodor, 186–87
Lyell, Charles, 230, 239–40
Lynx, 71

M

MacDonald, Ranald, 44, 216, 250
Mackenzie, Alexander (explorer), 144, 216, 223
Mackenzie River, 68
MacNee, Daniel, 179–80
Madeira, 27
Madrone, 122
Magnetic variation, 178, 192, 195
Magpie, 69
Malaria. *See* Intermittent fever
Manhattan Island (New York), 11
Manzanita, 122
Maple, 43, 46, 50
  vine, 46, 123
Mariposa lily. *See* Lily
Mark Hatfield Wilderness Area, 61
Mats (rush and cattail), 51, 66, 102, 114, 137
Maui, 241
Mauna Kea, 229, 230–33, 239, 241, 243
Mauna Loa, 232, 236–38, 239, 240, 243
McAbee fossil site, 222
McDonald, Archibald, 68, 88, 248
  and Douglas 44, 102–4, 106, 112, 145, 146, 211, 216, 224, 226
  tribute to Douglas, 250
McDonald, Finan, 44, 71
  and Douglas, 57–59, 101–2, 106–9, 120, 154–55
  attacked by buffalo, 156–58
  family, 102
McDonald, John (Finan's brother), 160

McIntyre Bluff, 220
McKay, Jean Baptiste, 71, 101–2, 120–27, 133
McKenzie, Alexander (HBC clerk), 39, 67, 194, 197
McKenzie, Donald, 160–62
McLeod, Alexander Roderick, 68, 71, 117–27, 133, 143–47, 197–98
McLeod, John, 44, 49, 74, 75, 76, 77, 78
McLouglin, John, 42, 44, 45, 65, 109, 188, 193, 197, 199, 227
  and Douglas, 42–45, 69, 71, 73, 74, 79, 94, 102, 115, 117–18, 132, 143–47, 197, 201
McLoughlin, Marguerite, 44
Menzies, Archibald, 26, 38, 41, 47, 63, 169, 173, 175, 231
Menzies Island, 47, 63
Menziesia, 124
Mercurous chloride. *See* Calomel
Mercury, 129, 196
Mersey River, 2
Methow River, 112
Mexico, 201, 207. *See also* California
Mica Dam, 150
Michaux, Andre, 25
Michigan Territory, 14, 17
Milkvetch, 174
Minerals, collected by Douglas, 50, 88, 145, 176
Minerva (schooner), 241
Mint, 123
Mission Creek, 219
Missouii (plant), 87
Missouri River, 13
Mistletoe, 84
Moccasins, 110, 145, 150, 216
Mohawk River, 14
Molokai Island, 241
Molasses, 38
Monkeyflower, 68, 106, 174
Monte Lake, 220
Monterey, California, 169, 200, 201–4, 206–8
Moonwort, 17
Moose, 155, 156

Mosquitoes, 97, 102, 106, 113, 160
Moss, 69, 74, 76, 78, 105, 192, 200, 215, 231, 233
Mother Carey's chickens. *See* Storm petrels
Mount Brown, xi, 151, 188
Mount Hood, 43, 61, 198, 227
Mount Hooker, 188
Mount Jefferson, 43, 120, 197
Mount Parrii (Oahu), 192
Mount St. Helens, 43, 67, 227
Mountain beaver, 141–42, 146, 176
Mountain goat, 146, 149, 175
Mouton gris. See Bighorn sheep
Mulberry bark, 236
Mullet, 234
Multnomah Falls, 61
Munro's Fountain, 104
Murray, John, 171
Music. *See* Dancing, Flute, Singing, Violin
Musket. *See* Firearms
Musket balls, 113, 130, 137, 149, 157
Mustard (plant), 143
Myrtle. *See* California laurel

N

Nachako River, 224
Nelson River, 163
Neptune, 28
Nespelem Canyon, 78, 145
Nets (fishing), 43, 50, 127
Newspaper, 123
Newt, western, 41
New Caledonia District, 93, 216–17, 218–26, 229
New York City, 10, 11, 13, 20
New York Horticultural Society, 14
Nez Perce Indians, 103, 105, 106
Niagara Falls, 17
Nightcap, x, 109, 149
Nimrod (ship), 21, 23
Ninebark, 84, 119, 122
Nipissing guide, 154
Northern lights, 163
Northport, Washington, 147

North Thompson River, 220
Northwest Territory, 68
North West Company, 42, 44, 76, 138
Norway House, 159, 162
Nose, The (landmark), 98
Nuttall, Thomas, 19, 20, 25, 161, 172, 214
  Genera of American Plants, 19, 53
  Nymphoides, 2

O

Oahu, 192, 208, 230, 240
Oak Point, 139
Oaks, 9, 14, 15, 16, 25, 118, 121, 132, 159
Oats, 193
Obsidian, 238
Oilcloth, x, 54, 80, 146, 151
Okanagan Lake, 220
Okanogan River, 78, 171, 218–19
Old Scone, 3
Olla Piska, 47
Olympia, Washington, 216
Omak, Washington, 219
Onion, 85, 87, 253
Opium, 34, 86, 239
Orchards, 9, 10, 12
Oregon grape, 24, 122
Oregon myrtle. *See* California laurel
Orchids, 18, 101
  coralroot, 18
  ladyslipper, 16, 18, 101
  pogonia, 18
  showy orchis, 18
  white bog, 48
Osoyoos Lake, 219–20
Otter, sea, 213
Owyhee (ship), 193, 200

P

Pacific Ocean, 31–34, 38, 54, 192, 201, 223
Palouse River, 106
Palus Indians, 103–6
Paper, plant pressing, 7, 40, 48, 53, 74, 79, 80, 82, 99, 112, 118, 124, 155, 235
Parsley, desert. *See Lomatium*
Pass Creek, 122–23

Peace River, 68
Peach, 11, 15
Peale's Museum, 13, 80
Pear, 9, 11, 24
Peas, 44, 193
Pelican, 53, 159
Pemmican, 97
Pend Oreille River, 147
Penstemon, 52, 93, 174
Peony, Asian, 99
  Brown's, 98–99, 155, 195, 253, 255
Persoon, Christiaan Hendrick, 16
Perth, Scotland, 3, 4, 5
Petrel, 28, 30, 54
Petroglyph, 52
Petticoat, 125
Philadelphia, 13, 19–20
Phlox, 52, 53, 79, 93
Pickering, Charles, 247–48
Pictograph, 52
Pigeon, band-tailed, 58, 66
  domestic, 21
  passenger, 160
Pine, 78, 133, 145, 173, 207
  digger, 203
  lodgepole, 80, 222
  Monterey, 252
  ponderosa, 80, 83, 84, 106, 220
  sugar, 59, 71, 117, 120, 125, 127–31,
    147, 168–69
  western white, 147, 149
Pinedrops, 11, 18
Pinus lambertiana. See Pine, sugar
Pinus sabiniana. See Pine, digger
Pipe, tobacco, 9, 46, 65, 70, 97, 103, 112,
  130
Pipestem, ceremonial, 155
Pistol. *See* Firearms
Pitcher plant, 2, 15, 20
Plant press, 7, 48, 53, 77, 81, 101
Plum, 11, 12–13, 24
Pondweed, 106
Poplar, 223
Potatoes, 38, 44, 137, 151
Potlatch, 196
Pots, 31, 68, 112

Prairie chicken, 182
Presbyterian, 119, 146, 161
Preserving powder, 124
  See also Specimens, Taxidermy
Preston, Robert, 5, 6
Pretty (Kalapuya chief). *See* Tochty
Priest Rapids, 77, 93, 112–13, 144, 171, 226
Prince, William, 11–12, 21, 24
Prince of Wales (ship), 164–65, 167
Princess Raven, 44
Principles of Geology, 230
Pronghorn antelope, 156
Provenchier, Bishop, 162
Ptarmigan, 182
Puget Sound, 215–16
Pumpkin, 234
Pursh, Frederick, 25, 61
  Flora Americae, 26, 54
Pyrola, 11, 23

Q
Quail, 21, 122, 181
Quamash. See Camas
Quarantine, 10
Quesnel River, 223
Quercus. See Oaks

R
Raccoon-skin wrist guard, 129, 131
Raft, 125, 128, 154
Raspberry, 42, 46, 231, 248
Rat, bushy-tailed wood, 96, 102
Rattlesnake, 93
Raven, 69
Red Coat (guide), 113
Red River colony, 158, 159, 160–62
Redwood, 203
Reed grass, 119
Reflecting circle, 246
Regent Street, London, x, 167, 183, 210
  See also London Horticultural Society
Reindeer. *See* Caribou
Rheumatism, 18, 86, 100
Rice, 128, 132, 133, 134, 216
Richardson, John, 25, 26, 52, 68, 152,
  158, 176, 188–89

Rifle. *See* Firearms
Rio de Janeiro, 28–29
Rio Plato, 30
Robado (Spokane guide), 110–12
Roberts, George, 214–15, 220–21
Robinson, Alfred, 203
Robinson Crusoe, 5, 31, 32
Rochester, New York, 14, 18
Rock Island Rapids, 93, 112
Rocky Mountains, ix–xiii, 13, 73, 148–53, 161, 180, 182, 222, 254
Rome, New York, 14
Roots, edible, 45, 46, 54, 56, 66, 79, 83, 87, 125, 134, 137, 171, 234
Rose 17,
Roseburg, Oregon, 129
Rum. *See* Alcoholic beverages
Rushes, 43, 51, 54
Russia, 186
Russian-American Fur Company, 68, 76, 206

S
Sabbath observance, 20, 119, 149, 161, 162, 236
Sabine, Edward, 15, 178, 185–87, 198, 216, 249
   correspondence with Douglas, 191–92, 215
   tribute to Douglas, 249–50
Sabine, Joseph, 8, 10, 16, 146, 183, 209
   and Douglas, 9, 24, 25, 52, 53, 63, 73, 168, 171–72, 174, 185, 190, 203, 209
   correspondence with Douglas, 93–94, 102, 159
   sister of, 169
Sagebrush, 51, 76, 78, 93, 95, 110, 144
Sahaptin tribes, 51
Salal, 38, 40, 41, 55, 63, 122, 123, 139, 163
Salamander, 41, 106
Salish tribes, 81, 146
Salmon, 60, 65, 70, 93, 95, 105, 114, 129, 206
   dried, 38, 97, 133, 139

fishery, 43, 46, 50, 52, 81, 107–8
   salted, 201, 214
Salmon River, 220
Salmon trout. *See* Steelhead
Salmonberry, 38
San Francisco Bay, 205, 229
Sandstone, 50
Sandwich, Ontario, 16
Sandwich Islands. *See* Hawaiian Islands
San Juan Bautista, California, 203
San Luis Obispo, California, 205
Sanpoil River, 92
San Rafael, California, 205
Santa Barbara, California, 205
Santa Clara, California, 203
Santa Lucia Mountains, 207
Santiam River, 136, 198
Saprophyte, 15, 16, 18
Sarsaparilla, 152
Saskatchewan River, 154–58
Sausalito, California, 229
Sawmill, 193, 231
Schachanaway, Cowlitz chief, 67
Schuykill River, 19
Scone, 3–5, 159, 179
Scone Palace, 4, 9, 175, 249
Scotland, 10, 213
   Douglas and, 3–6, 27, 56, 77, 178–80, 233, 241
Scottish fur traders, 44, 83, 138, 160–62, 220, 254
Scottish motto, 210
Scottish terrier. *See* Billy.
Scouler, John, 27–35, 37–42, 47–48, 50, 62, 63, 175–76, 213–14, 241
   correspondence with Douglas, 71, 77
Seaman's Chapel, Honolulu, 241
Seashells, 30, 38, 39, 138, 230, 236
Seaweed, 28, 30, 56, 138, 213, 215
Sedge, 51, 54, 56, 67, 142, 159
Sequoia, 203
Sewelel. See Mountain beaver
Sextant, 187
Sheep, 231
Sheep Creek, 147
Shooting star (flower), 79

Shoshone Indians, 100, 195
Shrub-steppe ecosystem, 51–52, 76–80, 95, 110–11, 144
Siberia, 187, 229
Silversword, 231
Simpson, George, 78–79, 138, 159, 161, 163
Sinbad the Sailor, 5
Singing, 92, 104, 144, 148
Sinixt. *See* Lakes Indians
Sitka, 206, 207, 224
Skeena River, 224
Smallpox, 10
Smith, Jedediah, 198
Snake Indians. *See* Shoshone Indians
Snake River, 45, 77, 93, 103, 140, 161, 180, 195
Snare, 127
Snipe, 84
Snow bunting, 151
Snowshoes, x, xi, 148, 151, 152
Soap Lake, 219
Songs. *See* Singing
Sonoma, 205
Sourwood, 20
Spatterdock, 106
Spear, fishing, 43, 108, 127, 132
    weapon, 130
Specimens, animal, 48, 57, 58, 63, 70, 85, 96, 102, 113, 120, 141, 146, 174–75, 176, 213
    bird, x, 30, 33, 48, 58, 63, 92, 140, 145–46, 159, 162, 174–75, 176
Spirea, 56
Spleenwort, 233
Spokane Falls, 107
Spokane Garry, 161
Spokane House, 83–85, 107, 161
Spokane River, 78–80, 83–85, 92, 98, 106–8, 110
Spokane Indians, 85, 106–8, 110, 161
Spokane, Washington, 107
Spring beauty, 81, 146, 220
Spruce, 153, 222
    Englemann, 80, 149
    Sitka, 124, 252

Spurge, 46
Squash, 234
Squirrel, 95, 102
Starfish, 30
Staten Island, 10
Steelhead trout, 80, 126, 127, 128, 134
Stikine River, 68
Stockings, 32, 74, 109, 149, 150, 232, 255
Strawberry, 231
Stuart, John, 154, 156
Stuart Lake, 224
Sturgeon, 38, 43, 55, 69, 139
Sturgeon River, 154
Sturgeon-nosed canoe. *See* Canoes
Suckers, 47, 148
Sugar, 60, 97, 112, 123, 201, 216
Sugarcane, 231, 234
Sulfur, 111, 235
Surveying instruments, 186–87, 191–92, 194–95, 197, 201, 204, 210, 215, 231, 236
    See also individual instruments
Swallows, 86, 143
Sweat lodge, 47
Sweet potato, 234
Synopsis Plantarum, 16

T
Table Mountain, 60
Tamarack. *See* Larch
Taro, 234
Tarweed, 119
Taxidermy, x, 26, 48, 57, 85, 92, 140, 145, 159, 175, 176
    See also Specimens
Taxonomy, 7, 19, 53, 79, 173–74, 252
Tea, 54, 60, 65, 66, 86, 97, 98, 100, 110, 112, 123, 125, 132, 133, 135, 137, 201, 216, 239
Teal, 32
Telescope, 237
Tent, 43, 48, 59, 76, 80, 86, 105, 118, 128, 133, 135, 139, 146, 216, 229, 236
Tern, sooty, 32
Tha-mux-i (Chinook elder), 65–67
Thames River, 23

The Beard. *See* Tha-mux-i
Thermometer, 48, 52, 87, 93, 187, 234
Thimbleberry. *See* Raspberry
Thistle, 120
Thomas, David, 17
Thompson, David, 44, 80, 83, 108, 150, 188
Thompson Rapids, 80, 91
Thompson River, 220–21
Tobacco, 4, 9, 15
  native species, 58–59
  smoking, 59, 65, 97, 112, 113, 114, 124, 128, 129, 130
  trade, 50, 56, 58, 65, 67, 75, 84, 97, 102, 106, 110, 113, 114, 118, 121, 125, 217
Tochty (Kalapuya chief), 137
Tollgate, Oregon, 99
Tolmie, William, 227
Tongue Point, 40, 65
Torrey, John, 10–11
Tortoise, 32
Toupin (Nez Perce interpreter), 105
Trade items, 56, 62, 67, 125, 139, 217
Trillium, 17, 81
Trinidad, 2
Trinket. *See* Trade items
Tropic of Capricorn, 30
Tropicbird, red–billed, 28
Trunk, 48, 109, 245
Tule. *See* Mats
Tumtum, Washington, 83
Turnips, 49
Turtle, sea, 27, 32
Twinflower, 151

U
Umatilla River, 99, 100, 195
Umatilla Indians, 144
Umbrella, 237
Umpqua Mountains, 70, 120, 125, 127–32, 169
Umpqua River region, 59, 71, 117, 122–134, 197, 198, 205
Umpqua Indians, 125, 126–33, 169, 198
University of Pennsylvania, 13, 19

United States Exploring Expedition, 247
Utica, New York, 14, 18

V
Valleyfield, Scotland, 5
Vancouver, George, 26, 231
Vasculum, x, xi, 12, 15, 16, 48
Velella vellela (jellyfish), 28
Venison, 122, 126, 128, 133, 137
Vetch, 122
Vigors, Nicholas, 178
Violin, 153, 156
Volcano, Hawaiian, 208, 230, 239–40
Voyageur, 45, 46, 50, 51, 74, 80, 92, 133, 136, 144, 145, 147, 148, 151, 155, 160, 194

W
Walla Walla River, 76, 95, 97–98, 253
  North Fork, 101
  South Fork, 98
Walla Walla Indians, 95
Wallula Gap, 76
Wapato, 66
War of 1812, 15, 43
Wasp, 119
Watch, 106, 126, 245, 247
Water lily, 106
Weaving, 56, 109
Wenatchee, Washington, 93
West Point, New York, 14
Whaler's Harbor (Sausalito), 229
Wheat, 193
White Bluffs, 77, 93
"White Fleshers." *See* Grouse, spruce
Whitefish, 47, 148, 153
White Mountain. *See* Mauna Kea
Wildfire, 61
Willamette Falls, 57
Willamette River, 43, 57, 118
Willamette Valley, 57, 118–22, 135–37, 197–98, 227
Willapa Bay, 66, 139
William and Ann (ship), 26–35, 37–39, 42, 48, 49, 62, 63, 65, 188, 193
Williams Lake, 223

Willow, 43, 50, 107, 149
Wilson, James, 176, 215
Wintergreeen. *See* Pyrola
Woburn Abbey, 1
Wolverine, 151
Wolves, 78, 213
Women
    California, 208
    Chinook, 38, 56, 142
    and food preparation, 45, 46, 56,
        83–84, 87, 125
    Hawaii, 237
    Kalapuya, 137
    Salish, 87
    Spokane, 83–84
    tribal, 45, 46, 51, 125, 127, 254, 255
    Umpqua, 125, 127
    wives of traders 44, 45, 68, 83, 102,
        118, 120

Wood ducks, 21
Work, John, 79–80, 92, 93, 94, 102–9,
    110, 145, 146, 149, 175
Wormwood. *See* Antelope bitterbrush
Wrangell, Baron, 206, 207, 224
Wyeth, Nathaniel, 214

Y

Yakima River, 77
Yellowjackets, 119, 120
Yerba Buena Hill, 229
Yes (Cayuse guide), 48–49
Yew, Pacific, 46
York Factory, 163–65
Yukon River, 68

Z

Zoological Society. *See* London
    Zoological Society